Building Interactive Worlds in 3D

Virtual Sets and Pre-Visualization for Games, Film, and the Web

Supplementary Resources Disclaimer

Additional resources were previously made available for this title on CD. However, as CD has become a less accessible format, all resources have been moved to a more convenient online download option.

You can find these resources available here: www.routledge.com/9780240806228

Please note: Where this title mentions the associated disc, please use the downloadable resources instead.

Building Interactive Worlds in 3D

Virtual Sets and Pre-Visualization for Games, Film, and the Web

Jean-Marc Gauthier

Routledge
Taylor & Francis Group

LONDON AND NEW YORK

First published 2005 by Focal Press

Published 2017 by Routledge
2 Park Square, Milton Park, Abingdon, Oxon OX14 4RN
711 Third Avenue, New York, NY 10017, USA

First issued in hardback 2017

Routledge is an imprint of the Taylor & Francis Group, an informa business

Library of Congress Cataloging-in-Publication Data
Gauthier,Jean-Marc,1960–
    Building interactive worlds in 3D :pre-visualization for games, film, and the Web /
Jean-Marc Gauthier.
     p.cm.
Includes bibliographical references and index.
ISBN 0-240-80622-0 (alk. paper)
1. Computer games—Programming. 2. Three-dimensional display systems. 3. Virtual reality. I. Title.
QA76.76.C672.G369 2005
794.8'16693—dc22

                2004061933

British Library Cataloguing in Publication Data
A catalogue record for this book is available from the British Library

ISBN 13: 978-1-138-40334-5 (hbk)
ISBN 13: 978-0-240-80622-8 (pbk)

Cover Design: Eric DeCicco
Composition: SNP Best-set Typesetter Ltd., Hong Kong

Table of Contents

Chapter 3

Interactive Textures and Lighting Design.................................. 83

Chapter 4

Chapter 5

Chapter 6

Chapter 7

Chapter 8

Acknowledgments

After writing the last page of this book, I feel the same deep feeling of satisfaction that I experience when coming back home after an amazing trip. When I look back at the manuscript, I realize how much the process of writing this book was shared with many people that I would like to thank.

Thanks to my loving Hillary and Isabelle (Gauthier) who are my best advisers, coaches, reviewers, and supporters.

Thanks to several explorers of virtual spaces at ITP:

to Miro (Kirov) for his passion and his kindness,

to James (Tunick) for his out-of-this-world enthusiasm and talents to organize and communicate,

to Zach (Rosen) for asking the right questions and bringing amazing answers,

Thanks to Noa (Steimatsky) for her friendship,

to Amy Jollymore who nurtured the writing process and to Diane Cerra who introduced me to Amy,

to my students at ITP, Tisch School of the Arts, New York University, who tested the demos and influenced the presentation of many topics of this book,

to Ken Perlin, Fabien Bariti, Mike Olson, Eric Socolofsky, Antoine Schmitt, Ronald Hof, and Adrian Siminovitch, for their writings, their code, plug-ins, and expertise,

to Morekwe Molefe, Cara Anderson, Phil Bugeau, Bertrand Duplat, and Virgile Delporte, who helped with the making of the book and the companion CD-ROM,

to Florent Aziosmanoff, Le Cube, Michel Hamelin, Antenne 2, Martin Hash, Animation Master—Hash Inc., Thierry Martin, Bernard Morel, Ville de Nice, Martin Nachbar, Advanced Educational Services—School of Medicine—NYU, Christopher Ratte, Institute of Fine Arts—NYU, and to clients, collectors, curators, journalists, and collaborators who created opportunities, gave their support, rolled the drums, and to those who became friends,

to the people that came with me on this journey, and special thanks to the new readers of this book!

Contributors

Jean-Marc Gauthier, Architect D.P.L.G., Paris; M.P.S., New York University. Jean-Marc, an architect from France, is enjoying a significant freelance career as an interactive designer, code programmer, and new media consultant. His background is as technically oriented as it is rooted in design. JMG has been involved in virtual worlds since 1992, when he created "The Liquid Map," an award winning real-time mapping simulator for decision makers. JMG has created virtual space installations inside museums and a low-cost, multiple-screen display for educational purposes, such as the Dynamic Virtual Patient and the Aphrodisias Project. JMG has designed several urban installations inside the city—for example, Nighthawks allows viewers to use cellular phones in order to control virtual spaces projected on a building.

JMG's installations have been presented in museums and international conferences including the MAMAC (Nice), Chelsea art Museum (New York), American Museum of Moving Images (Queens), Massachusetts Institute of Technology's International Conference on Design Computing and Cognition (Cambridge), Festival 1ier Contact (Issy-les-Moulineaux), Villette-Numerique (Paris) and Siggraph.

JMG is the founder of www.tinkering.net, a consulting agency specialized in designing real-time 3D content for medicine, archeology, fashion, urban installations, virtual sets, and events. JMG has created interactive designs for New York University's College of Medicine and Institute of Fine Arts, the Cousteau Society, the Institute for Environmental Studies at the University of Houston, the Four Oaks Foundation, USANetworks, the Academy for Educational Development, La Cite des Sciences et de l'Industrie (Paris), and MAMAC (Nice). JMG is the author of several books and essays including this book and *Creating Interactive 3D Actors and their Worlds* (2001), from Morgan Kaufmann Publishing.

Florent Aziosmanoff, new media artist, has a Masters in Psychosociology and Cognitive Sciences from the University of Vincennes, Paris 8, France. In 1988, he created a nonprofit art and new media association, ART3000, and today runs the Festival Premier Contact, a city-wide installation of interactive electronic art, with his brother Nils. He is currently Art

Director at the Cube in Issy-les-Moulineaux, France, where he directs the "Digital Art Atelier," open to multimedia authors working on projects using virtual reality and real-time interactive systems (artificial intelligence and artificial life).

Miro Kirov, born in Sofia, Bulgaria in 1962, graduated with a BFA, MFA in sculpture from High School of Fine Arts and The Academy of Fine Arts in 1989. In 1990, he received a bronze medal award from the Prestigious International Bienalle in Ravenna, Italy. In 1990, he moved to New York and took part of group shows including CB's Gallery (New York), Roxbury Gallery (Roxbury, Connecticut), and Elsa Mott Ives Gallery (New York). His solo exhibits are displayed in Hemus Gallery in Sofia, Katelbach Gallery in Hamburg, Germany, and Werkstatt Gallery in Gelsenkirchen, Germany. In 2000, Miro was hired by the New York University School of Medicine and he is currently working on his MPS at the university's Interactive Telecommunication Program. His interests are in 3D interactive and immersive environments, 3D animation and modeling, video, and conventional fine arts.

Ken Perlin is a Professor in the Department of Computer Science at New York University. He is the Director of the Media Research Laboratory and the Co-Director of the New York University Center for Advanced Technology. His research interests include graphics, animation, and multimedia. In January 2004 he was the featured artist at the Whitney Museum of American Art. In 2002 he received the New York Mayor's Award for Excellence in Science and Technology and the Sokol Award for Outstanding Science Faculty at New York University. In 1997 he won an Academy Award for Technical Achievement from the Academy of Motion Picture Arts and Sciences for his *noise* and *turbulence* procedural texturing techniques, which are widely used in feature films and television. In 1991 he received a Presidential Young Investigator Award from the National Science Foundation.

Dr. Perlin received his Ph.D. in Computer Science from New York University in 1986, and a B.A. in theoretical mathematics from Harvard University in 1979. He was Head of Software Development at R/GREENBERG Associates in New York from 1984 through 1987. Prior to that, from 1979 to 1984, he was the System Architect for computer generated animation at Mathematical Applications Group, Inc., Elmsford, NY, where the first feature film he worked on was *TRON*. He has served on the Board of Directors of the New York chapter of ACM/SIGGRAPH, and currently serves on the Board of Directors of the New York Software Industry Association.

Zach Rosen is a Resident Researcher at the Interactive Telecommunications Program (ITP) at New York University. Self-taught in programming and graphics, he majored in computer science and minored in film at Brandeis University, where he pursued interests in artificial intelligence and human–computer interaction. Zach went on to get his M.P.S. from ITP where he wrote and programmed his thesis on extending the capacity of massively multiplayer online worlds using distributed client-server clusters. His recent projects include programming for the 3D re-creation of an archeological site, developing 3D shaders and other technologies for a medical grant, and designing, writing, and publishing Bluetooth software for cellular phones.

Mike Olson graduated from Harvard University in 1996. He is a former research engineer at voice recognition pioneer Dragon Systems and Vice President of Technology at Affinity Financial. He is currently pursuing a masters degree at New York University's Interactive Technology Program focusing on new interfaces for virtual environments. His primary interest lies in creatively applying novel technology in the entertainment domain, including toy design and gaming.

Eric Socolofsky (http://transmote.com) wanders the gray areas between architecture, interactive design, and digital and physical art. After working as an architect for a time in Chicago, he moved to Brooklyn, New York and began experimenting with interactive installations at the Interactive Telecommunications Program at New York University. He is currently further developing concepts of interactive space in his residency at Eyebeam's Production Artist Studio in Chelsea, New York. He occupies his spare time designing and programming user interfaces and games, and teaching the occasional workshop or class.

Antoine Schmitt atypically first lead a successful software designer career before becoming an active artist, exhibited internationally. For 25 years, Antoine Schmitt has been designing human-computer interaction software in various original companies in Paris and California (Act, NeXT). As an artist and programmer, Antoine Schmitt stands at the crossing of abstraction and dynamic simulation. He uses programming as a first-class material to create installations, online exhibitions, performances, and CD-ROMs.

This work has been awarded in many international festivals. As a curator, member of jury, speaker, and editor of the gratin.org portal, Antoine Schmitt explores the field of programmed art. Antoine Schmitt lives and works in Paris.

Full artist biography: www.gratin.org/as/

Software: www.as-ci.net

Noa Steimatsky is Assistant Professor of the History of Art and Film Studies at Yale University. She has just completed her book on landscapes in Italian film. Her research engages with questions of cinematic realism and modernism, poetic figuration, and documentary practices in the postwar era. She has won the Getty postdoctoral grant for her research on the face in film and, most recently, the Rome Prize for her work on Italian cinema in the 1940s.

James Tunick is a new media engineer, artist, and founder of Studio IMC, a new media design studio and artist management agency (www. StudioIMC.com). He has worked with clients such as the Museum of Modern Art PS1, Hennessy, Diesel, and others. Tunick is currently developing immersive and interactive visual displays with Studio IMC partners Tony Rizzaro and Brian Karwosky for use in retail stores, architecture, museums, and schools. In addition, he is developing interactive environments and audiovisual installations for use on stage and in architecture with Jean-Marc Gauthier, Studio IMC Principal Designer and Architect.

Among the many multimedia events Tunick has produced and organized is Interfaced Culture at Yale University, an international conference on interactive new media technology in the arts for which he received a grant with Yale professors Dr. Mathew Suttor and Dr. Kathryn Alexander in 2002. He also curated a Studio IMC show at the Chelsea Art Museum in 2004 called "Convergence: The Collision of Physical and Virtual Space in Digital Art," featuring the works of Studio IMC's top artists and engineers.

Tunick graduated from Yale University in 2003 and is currently earning a masters degree in new media at New York University's Interactive Telecommunications Program. In addition, he organized the 2nd Annual Studio IMC Expo tradeshow in New York in April, 2004, launching a Studio IMC magazine about technology, politics, and culture to be launched in 2005, and raising VC funding for the Studio IMC Research Lab & Gallery Space to be founded in New York in December 2006.

Preface

Dear J-M,

Hope your visit in Paris, our native town, is good. I recall the time, back in the late 1980s, when I visited you there in rue Lalande, and a chat over your plate of choucroute in La Coupole.

I am reading the chapters you sent me, and they make me ask myself some old questions in new ways: on optics and mental processes which are so often used as metaphors for each other; on how cinema's modes of articulation—camera movement, framing, editing—inform us about viewing habits generally, and about how those may be subverted.

I have never had any experience with VR so I was not always sure about our "object" of conversation here—I tried to imagine it in my own traditional world. What your vivid settings allow me to do is imagine it through cinema. For this I am grateful. Naturally, seeing cinema as a threshold to VR should be historically sound and pedagogically valuable. I am fascinated by the mention of Kubrick's *2001* several times in your introductory chapter. Since I am so attached to the photographic basis of cinema, I kept in mind the simple fact that *2001* is strictly *pre*-digital effects, that its production involved massive technologies and machineries and labor to achieve its astonishing images: those smooth motions, that spatial depth and absorption. I believe that, compounding the scope and enhanced resolution of a 70 mm negative, *2001* was originally designed to be screened on tilted screens that "embraced" the viewer's field of vision.

I was reminded of the material weight of all these pre-digital effects just last week. I was in Bologna where *2001* had just been screened in the Piazza Maggiore, along with such other extravagant productions such as Tati's *Playtime* (almost contemporary with the Kubrick, in fact). I was in the Cineteca film archive the following week doing my own research while they were packing the films to be shipped back to various places. And there they were: the actual, numerous, 70 mm reels of *2001*, each packed in its thick metal case with a carrying handle. I grabbed one of those precious reels and tried to pick it up, but couldn't. *It was too heavy.* This is an important lesson about that older technology, cinema, and it came to mind as I was reading your chapters.

So I am now compelled to ask: What are the material imperatives of VR? It may not claim such weight and bulk, but how *does* it claim a space? What position does the viewer, and the viewing device (itself simulated) occupy therein? How does it constitute a position, a presence, even an identity of sorts? Your tracing of a genealogy for virtual cameras in some great moments of modern cinema suggests to me that, across different media, and in both theory and practice, the two of us coming from different fields may be exploring similar terrains. I know that these are also the concerns that haunt our most adventurous students. What does it mean that a camera intrudes upon a scene, that its presence confronts our anthropomorphic habits of looking and sensing our environment, confronting as well our emotional responses, our relationships with people and objects and places?

Reproducing with new media the traditional forms and values of mainstream cinema may be a good lesson but, reading your account of the positioning of producer and consumer, author and viewer in VR suggests that surely there is something new, possibilities yet to be imagined and pursued in radically new ways through the limitless choices opened up in VR. For "choices" cannot remain a mechanical navigation akin to those maddening telephone menus that end up impoverishing experience. We seek rich and compelling relationships in every sense—with the natural and the built environment, with things, with each other, and with earlier instances of experience, with time. To my cinematic sensibility, these ideas emerge in your evocation of the "gaps" that open up in the visible arena of Hitchcock's *Rear Window*, in the obstructions of the wild environment of Antonioni's *L'avventura*. How would such a compelling, difficult space—a space of play, desire, mystery, and anxiety for the mind as much as for the eye—translate to VR? What would it mean in that virtual domain that (as Antonioni observed) people disappear every time they leave the room? Perhaps in that virtual *there* this would be completely reasonable!

Shall we look at the island sequence together when we meet again?

Yours,

Noa

late July 2004, New York

Noa Steimatsky teaches film studies in the Department of the History of Art at Yale University.

Introduction

I have found ways to express myself as a writer, moviemaker, photographer, and architect but still felt the need to communicate ideas that did not fit into any of these domains. New possibilities for expression emerge in the design of virtual spaces. Designing virtual spaces helps me convey ideas in ways that cannot be expressed the same way in writing, movies, photographs, architecture or other visual art. Surrealist artists from the 1920s (Man Ray, Marcel Duchamp, Louis Bunuel) felt that movies could tackle situations or emotions that could not be depicted in writing or in the traditional visual arts of their times. For similar reasons, topics covered in this book stretch beyond the strict domain of virtual spaces. They relate to larger concepts like the future of cinema, the place of virtual spaces in tangible public spaces, defining new experiences for the viewer of a virtual space, or developing new types of associations between several media that may involve both the virtual and real. I hope that this book will help you to look at virtual spaces as an immense territory for experimentation where you can build, explore, and play more easily and faster than in the physical world.

Virtual sets are redefining the way film directors, dancers, scientists, medical researchers, architects, TV producers, and web designers can develop simulations of real world situations. Examples of immersive virtual sets described in this book range from the 3D pre-visualization of a movie scene prior to shooting on location, to a virtual visit to an agora of the antique city of Aphrodisias, Turkey. Applications of virtual sets I cover range from interactive TV games to information systems created for decision makers.

Sections of this book cover the creation of real-time interactions between an audience and a virtual actor using sensors and artificial intelligence. I explain how a viewer can interact with virtual sets using simple inputs such as live-video, gestures, or sound, as well as up-to-date technologies like cellular phones.

"The beauty of nature lies in detail; the message, in generality. Optimal appreciation demands both, and I know no better tactic than the illustration of exciting principals by well-chosen particulars." (From Stephen Jay Gould. *Wonderful Life: The Burgess Shale and the Nature of History*. New York: W. W. Norton & Company, 1989.)

Gould's ideas about how to describe the living environment of some of our planet's first animals can be a source of inspiration for someone who attempts to design virtual worlds. Like in Gould's first sentence, the designer's special attention to details conveys a sense of beauty to the viewer; but the vision of virtual spaces needs to be much broader.

This book addresses the creation of simple 3D interactive environments where some elements of virtual spaces such as interactive textures, virtual cameras, kinematics, real-time physics, and self-determined virtual actors can be found. Each chapter covers one aspect of interactive 3D, illustrated by a complete tutorial. I give in-depth explanations of reusable building blocks and modifiers that can be applied to 3D objects, cameras, lights, and animated characters. More than an introduction to the fascinating world of 3D, this book also offers insight into how to design real-time special effects for movies, TV broadcasts, interactive games, or even 3D games. Examples will utilize Virtools, Maya, and Lightwave software to show how to design and program virtual sets.

Two main directions of this book are to create interactivity and a sense of presence in virtual spaces. The place of the viewer is a key element of the design of a virtual space. The viewer is in motion inside a physical space where the expected notion of facing a screen tends to disappear. This book shows installations where viewers are invited to walk around installations where 3D content is presented on screens extending in multiple directions of a space, challenging the reactivity of the virtual space around them. The design of virtual cameras with artificial intelligence will also be addressed in order to create relationships between a virtual camera and the context of a scene.

THE ORGANIZATION OF THE BOOK

This book is organized like a "cookbook for interactive 3D," bringing the reader through a personal process. Each chapter introduces specific examples of virtual sets illustrated with hand-drawn sketches. The tutorials present step-by-step creation of virtual sets starting from early storyboards, notes, and sketches and evolving into fully working prototypes. This "cookbook" approach will help readers understand how they can better apply and then market their own skills. The companion CD-ROM includes files for the tutorials and a full trial—PC version of Virtools 3.0. You can play the examples inside the Internet Explorer window for Mac or PC. You can also recreate the tutorials in Virtools 3.0.

For more advanced readers, the book presents several applications of real-time physics that can greatly enhance the immersive experience in a virtual world. The book includes a collection of behaviors illustrating how to simulate natural phenomena such as virtual wind, fire, pyrotechnics, smoke, clothing, ropes, and hard and soft body collision.

- **Basic entries** (or keywords) for this book include *3D basic kit, animation basics, gravity, floors and stairs, live video, virtual cameras, interactive 3D lighting, communication environment, interactive character animation, interactive kinematics, broadcasting messages,* and *event-driven animations.*

- **Typical step-by-step tutorials** found in this book cover collision detection, virtual cameras inspired from movies, virtual characters and bots using path finding and terrain analysis, simulation of a crowd in motion, applications of virtual sets for movie sets, and artificial intelligence and gameplay for decision making.

- **More advanced entries** for this book include a virtual archeology project, architectural simulations, medical simulations, interactive kiosks, immersive displays in a museum, and urban installations. The book shows examples of virtual spaces created for research and interdisciplinary projects helping to communicate ideas to a group of diverse and highly creative people. These examples of virtual spaces were created in order to think about a problem in many different ways. The book covers the making of installations inside a museum gallery or an urban space. These examples show how viewers can break away from the traditional conventions of viewing art in a museum.

THE CONTENT OF THE CHAPTERS

Each of the following chapters covers a domain of virtual spaces.

- **Chapter 1** lays out the convergence of several artistic factors present in virtual worlds.

- **Chapter 2** presents the basic 3D kit with a tutorial on modeling and animation in Maya and the particular setups to export 3D content to Virtools.

- **Chapter 3** documents kinematics and all aspects of interactive motion. Topics covered are related to imitation of motion, lighting and physics, and the early steps of imitation of nature in virtual spaces where responses to situations are not predefined, but created on the fly.

- **Chapter 4** develops interactive textures.

- **Chapter 5** develops interactive path techniques presented both as key elements of storytelling and as examples of increasingly complex interactivity.

- **Chapters 6 and 7** cover virtual cameras with behaviors modeled on real world characters.

- **Chapter 8** probes the viewer's experience of virtual reality becoming an activity that can rival with some of the best cinematic experiences.

A BOOK ABOUT THE PROCESS OF CREATING VIRTUAL SPACES

The creative process and the production process are presented side by side through text, sketches of virtual worlds, and drawings created during the course of various projects. The emphasis given to the process of collaborative work helps the reader to understand that many different skills can be involved. Several interactive projects are documented in order to fully illustrate the connections between designing assets, programming digital art, and the viewer's experience. I hope that this effort of documenting the work in progress will help you to go beyond creative thresholds of artistic expression and invention.

The goal of a virtual space designer has become more and more about the imitation of living systems so other people can identify, evaluate, play, and create. Mapping internal states, mental desires, and perceptions seems to be the final frontier of virtual spaces and raises exciting issues of representation. Can we explain how to represent our dreams? How can the autonomous agents that you created represent their dreams? Whose dreams are they—yours or theirs?

I hope that you become more curious about your own ideas while reading this book and that you will enjoy discovering the making of virtual spaces.

CHAPTER 1
Emergence of Virtual Spaces

1   EMERGENCE

People living at the end of the nineteenth century saw the first movies. The Pathé brothers, owners of fun fair attractions and movie producers and distributors, presented movies directed by Melies to small audiences under a tent built next to other attractions such as magicians, fire throwers, and sword swallowers. The movie theater found its audience and its own specific space several decades after the first screenings of the Lumiére brothers, in a fancy restaurant called the Café de Paris. The layout of the movie theater, a space designed for a collective viewing experience, remains unchanged still today.

At the end of the twentieth century, online three-dimensional (3D) games reached audiences of millions of people sharing the same virtual space through a multiplayer online experience. The convergence of networking, the internet, and interactive 3D worlds changed a gaming experience based on one desktop into a virtual space experience played by an unlimited number of people.

Comparing the evolution of virtual spaces and the evolution of other media helps us understand the emergence of new types of viewer's experience and of new audiences through time. Trying to define an audience and a viewer's experience for virtual spaces seems to be a priority for a young medium with a direct appeal to large audiences. This chapter tries to answer several questions about virtual spaces and the future of the viewer's experience.

2   DISCUSSION ABOUT VIRTUAL SPACES

This chapter takes you through a virtual conversation that takes place in a group that includes designers, producers, developers, and 3D artists.

More detailed information on the following participants can be found in Biographies.

Ken Perlin is a professor in the Department of Computer Science at New York University. He is the director of the Media Research Laboratory and the co-director of the NYU Center for Advanced Technology. His research interests include graphics, animation, and multimedia. Ken won an Academy Award for Technical Achievement from the Academy of Motion Picture Arts and Sciences for his *noise* and *turbulence* procedural texturing techniques, which are widely used in feature films and television.

Miro Kirov is a sculptor and a 3D artist from Bulgaria who works and lives in New York. For the past 4 years Miro has been creating a virtual human body for Advanced Educational Systems at New York University. Miro and I collaborate on the Dynamic Virtual Patient project and on virtual space installations with James Tunick and Studio IMC.

Zach Rosen is a designer and developer who lives and works in New York. Zach is a researcher at the Interactive Telecommunications Program at New York University, where he graduated. Zach and I collaborate on several virtual reality projects funded by New York University. Related projects illustrated in this book include the Aphrodisias project and the Dynamic Virtual Patient project.

Florent Aziosmanoff studied psychosociology and cognitive sciences before becoming artistic director at the Cube in Issy-les-Moulineaux, France. He created the association ART3000 and the Festival Premier Contact with his brother Nils and the team at the Cube. Francois directs the "Atelier de creation" at the Cube, a workshop for artists and authors creating virtual reality projects and real-time interactive systems. Artists and authors can develop projects from scratch using decision-making techniques including artificial intelligence and artificial life.

2.1   Let's Start the Discussion

What are you looking for when entering a virtual space?

Ken

When I enter a virtual space, the most important thing that I look for is a feeling that I'm entering a world that will touch me emotionally, that I can immerse myself in. It's a very childlike feeling of wonder when it works.

Miro

I am looking first for the ground under my feet. If the world is a simulation, then the virtual environment is our reality. Knowing that the world is a three-dimensional abstract expression floating in space is just another option.

Zach

When entering a virtual space, my first goal is to learn the methods of manipulating the space. This can take the form of locomotion, cursor navigation, sound manipulation, etc. During this brief orientation, I am looking for visual or auditory feedback to let me know that my actions are causing a reaction. Now that I know the tools that are available, the second step is to explore these methods to determine their capabilities. If I know how to make the character walk forward, can I get him to walk stairs? If I whistle to change the color of the screen, what happens when I whistle louder or softer, higher or lower?

Once I am comfortable communicating with the space, I can begin discovering what the purpose of the world is. As essential as instant feedback was to help orientate me with the controls, I am now looking for reference points or clues to tell me where to go or what to do. Is there a location I must discover or a mission I must accomplish? Am I here only as an observer or am I an active participant? At this point I am in the hands of the developers to guide me toward their intended purpose for the space.

What is the importance of the viewer in the design of virtual spaces?

Ken

The viewer is the most important thing in the design of virtual spaces, just as the reader is the most important thing in the design of a novel.

Florent

We can be sure of one thing: the viewer alone cannot "turn" any reality into a virtual reality environment without the input of the author. Let's take the example of a visitor walking inside the park surrounding the Chateau of Versailles, a seventeenth century garden designed and created by Le Notre. The viewer is immersed inside a landscaping project inspired by a certain vision of the world at the time of King Louis the XIV. Although everyone can browse freely and at his or her own will in the alleys of the park, each visitor is constantly reminded of the rules created by Le Notre. For example, the timing of walks and transitions between various sections of the garden are rigorously designed. The visitor can feel the dramatic perspective of the space starting from the castle, going over the "green carpet" and the Grand Canal. The view of the horizon is blocked by a row of poplars from Italy and projected toward the sky. The visitor, immersed inside this grandiose and calm space, can see thousands of visitors following their activities quietly. One can feel how the layout of a space suggests a world united and pacified. This is certainly the world that Le Notre, the author, intended to design.

Zach

When designing a virtual space, every decision must incorporate a consideration for the resulting effect on the user's experience. While it may be interesting to explore visual phenomena or push technical boundaries, the reaction of the user should remain a primary concern. Successful interactive installations entice the user to participate by using visual clues to suggest that something will happen if the user touches the mouse, screen, etc. If a virtual space is too active when there is no user input, viewers will take less interest in interacting and become passive observers. A space that understands what the user is feeling will help to orientate the user and improve his or her understanding of the purpose of the space.

Miro

The importance of the viewer is essential for virtual spaces. The role of the viewer can be passive, observant, active, or interactive. The whole purpose of virtual space design is to achieve an immersive and interactive experience between the viewer and the world.

Can you describe a viewer's experience—situation or emotion—that you find specific to a virtual space? Can you compare the experience of a viewer inside a virtual space with an interactive moviegoer's experience?

Ken

I think that virtual reality spaces are very good for communicating awe and mystery, if they are properly done. I have never seen an emotionally effective interactive movie, so it's hard to compare the two.

Miro

Our mind is a funny thing. It is like sponge. It absorbs everything. I had this strange experience of being totally immersed in some 3D world and vice versa. After long hours of 3D game playing, the experience of elusiveness of the surrounding world is a kind of inversion of reality that can be a dangerous thing!

The moviegoer can immerse into the story, the drama, but has no control over it. He or she is a passive contemplator of it. It is very linear in time. The virtual space viewer has more of a choice over actions and what to see. It is very much like life. We are in a certain situation and we have to make a choice with its consequences.

Zach

If there are no apparent guidelines or goals, the ideas of the developer will not be successfully communicated to the audience. The most common reaction that I see with viewers of virtual spaces is confusion. Unlike film or television, virtual spaces are flexible at being active or passive creations. The message of a virtual space can be a lost process when viewers get confused. There is a very sharp learning curve that takes place when someone decides to participate in an interactive installation.

Florent

We can be tempted to look at virtual reality the same way we look at movies. The viewer is in both cases taken inside an immersive sight and sound system, with the addition of a mouse or keyboard in the case of a virtual reality environment. But we can grasp major differences between virtual reality and movies when looking at the way real-time systems change the way to deliver content and the structure of the content being delivered. Differences may be even stronger than between theater and cinema. In virtual reality, the viewer's experience goes beyond sequential storytelling. Let's take again the example of the garden surrounding the Chateau of Versailles. The viewer's experience of a virtual reality tour of the garden may be very different than watching a movie about the park of the Chateau of Versailles. In one case, moviegoers follow a point of view that can't be changed; in the other case, viewers are free to choose their viewpoint and the location of the camera and to create their own editing, etc. I am not sure if traditional movie editing can be used the same way in virtual reality.

The way content is delivered in movies and in virtual reality is so different that I am not even sure that they can be compared. There may be a way to reuse some formal experiences borrowed from movies—for example, framing, moving the camera, and mastering the tempo of a scene.

Can suspension of disbelief be part of the viewer's experience in virtual spaces? Is this something that you are looking for as a visitor, as a designer?

Ken

Yes. In fact, I think it is all about suspension of disbelief. That is the factor that allows us to make extremely nonlinear choices in design (focusing on some things and ignoring others), much as a filmmaker might focus on only certain characters and leave others sketchy.

Miro

It depends on if your virtual space is mimicking the sense of reality and perspective, if the meaning of the word "space" is perceived as a tangible form. The Cartesian coordinate system usually used for virtual environments should not necessarily be a way to present things from our imagination, but it can used as an aid.

How can people learn from virtual spaces?

Miro

Well, virtual spaces can enhance the way we see things and memorize valuable information on the reality of things. We still see the world as a flat pancake. Only at some point of the coast, looking at the sea, can we see that the world is round on the horizon and yet we believe it's round from our knowledge. In this context, we would be able to experience different kinds of perspectives of representation with the help of the technology—for example, we could perceive a patch of grass from the point of view of an insect.

Zach

It is essential to remember that realism and believability are two separate entities. Designing a realistic space will lock you into a constant comparison between your work and its real world counterpart. It is a thankless struggle because the real world will always seem more "real." Rather than focusing on realism, bring the viewer into a new realm. Do not be afraid to let them know that this is a different type of reality where some things may be recognizable and some things may seem disturbingly different. As long as you are consistent, the audience will believe every piece of your new world.

As a designer of virtual spaces, you are blessed with an audience that expects to be transported to a new reality. Treat your viewers as you might treat an out-of-town guest: you may want to show them how to get around or what is available for them to see or do. Your space can be as close to or as far from the real world as you decide and your audience will follow you there, but if you leave them without any reference point, any rules, or a motive, they will feel lost and retreat from your space.

Is it possible to use some linear content to tell a story inside a virtual space?

Florent

The question is open. I currently design *Le temps de l' amour*, a virtual reality fiction that uses linear content. The viewer can jump in the story but can't control the delivery of the content. The viewer can only control his or her viewpoint on the story. The viewer is placed in a situation similar to Wim Wenders' angels in *Wings of Desire*. My contribution to this project is to give full access to all the levels of the story for the viewer, including present, past, and future. I am to give not only an unbiased report about what happened inside a specific scene but also the subjective, conscious, and subconscious viewpoints of a character in the scene. This open system lets you change your point of view on the story rather than changing the elements of the story. Since my original idea was to mimic a movie, this experience gave me the opportunity to think about how to create stories for movies by allowing the viewer to have total freedom to choose a viewpoint.

In movies, the director chooses a point of view for you. For example, the director chooses the main character or two main characters if there is a duo. Other secondary characters—who tell us other ways to see the story—help to structure, to contradict, or to embellish the main character. Even movies built around the opposition between two protagonists (such as a man and a woman or a cop and an outlaw) need to emphasize the point of view of one dominant character. This constraint comes from the limited duration of a film, which is about 2 hours. This is just enough time to present one way of looking at the story; using the unit of the sequence gives a way for the viewer to measure time, emotion, and cognition. We know that the cinematographic language is partly based on the ability of the viewer to project himself or herself on the main character. It seems impossible to change the main character, the viewer's referential, during the time of a sequence.

Are you interested in the utopian dimension of virtual spaces? Do you find a place for ambiguity, unexpected events, or failure in virtual spaces?

Ken

I don't see how you could have a utopia without ambiguity or unexpected events or failure. After all, those qualities are a necessary part of being human—without them we are not in a utopia but rather in a sterile lifeless place.

Miro

No, I do not believe in a utopian dimension of virtual spaces. The surprise event is crucial in this type of environment. It can be a great educational tool to teach people to see and react to. Unlike life, in virtual spaces failures can be fixed, conditions restored, resources regenerated. Virtual worlds are more forgiving in that sense.

Zach

The possibilities for expression in virtual spaces are truly endless. It is a medium that can combine nearly every other type of media and contains all of the flexibility and power therein. Utopias are one possibility, but dystopias are also effective as political or societal commentary or explorations of phobias. Ambiguity and discontinuity can also be used as effective tools as long as they are used intelligently and do not confuse the audience so much that they lose interest.

Is it possible to create storytelling systems with autonomous characters in virtual reality?

Florent

In 2001, I created an adaptation of the tale of *Little Red Riding Hood* with three autonomous robots.[1] The robots were small animated dogs with 20 motors, several sensors, cameras, and microphones. These truly autonomous machines could manage their environment and could acquire complex behaviors similar to artificial life and artificial intelligence. I was working on the simple story of *Little Red Riding Hood* in which the little girl, the wolf, and the hunter are set up inside a repetitive system of relationships. I was interested in the moral, social, and psychoanalytic aspects of the tale that can be found in the study done by Bruno Bettelheim.[2]

The wolf is looking for Little Red Riding Hood, who listens anxiously until the hunter brings the situation to an end. After the encounter, each actor is reset to walk around until a new set of circumstances brings the actors together again. The play is designed to repeat itself indefinitely with various versions of the same story line. The robots were created to play among people in the streets. They were designed to respond to an environment that could change according to the context—for example, interactions with viewers could interfere with the story line. Viewers could help to end the story, to protect Little Red Riding Hood, or to create new situations. The wolf looking for the color red could be misled sometimes by a kid's red shoes or a person's red handbag.

Creating this piece followed the same constraints that we apply to virtual actors inside virtual reality; the tools are also very similar. This piece required tweaking parameters for various behaviors and creating responses to external stimulations while controlling the equilibrium of the whole system. For example, if Little Red Riding Hood walks too fast, the wolf may never catch up with her. The dangerous wolf becomes a pathetic character, and the story line is completely changed.

---

[1] Bettelheim, B. (1976) *Psychanalyse des contes de fees—The Uses of Enchantment*. Paris: R. Laffont.

[2] Sony's Aibo robots created for 1erContact festival, organized by ART3000—Le Cube, Issy-les-Moulineaux (France). Urban installations of digital art are presented throughout the city.

I was very surprised to see how the play performed remarkably well on the functional level, on the storytelling level, and in relationship with the audience. I noticed that the audience did not want to interact directly with the piece. Children wanted to have a specific relationship with one of the characters and specifically with Little Red Riding Hood. Some of them tried to create a personal relationship with the character of their choice, trying to protect it from other actors or from other visitors. Watching these behaviors made me understand the kind of new space I was looking for.

Which references, books, movies, music, and examples of other topics would you recommend to the reader?

Ken

Turn off the lights, get a good projector with a good sound system, and go back and watch *2001: A Space Odyssey*. Think about the way Kubrick plays with time to create mystery, the way the film raises questions in your mind just by juxtaposing virtual spaces that are very different in our emotional and cultural landscape.

Miro

Alice in Wonderland, *The Little Prince*, Kafka, Joseph Campbell's lectures. Escher always fascinated me, Calder's sculptures, Van Gogh's skies. Some of the performances of Cirque du Soleil, Aida at the Met, Frank Lloyd Wright's Guggenheim museum building. For films, Fritz Lang's *Metropolis* is a classic; Terry Gilliam's *Brazil* too. Kubrick's *2001: A Space Odyssey* is not taking place in 2004. I find it fascinating to live through times that others dreamt about in the past. Spielbergs' futuristic experiments with today's cinematic story telling are interesting too.

In *Microcosmos*, the movie, a patch of grass is shown from the point of view of an insect. What a wonderful example of how to see the world in a different perspective.

2.2   Notes on the Discussion

I added the following notes after "listening" to the discussion.

Designing virtual spaces is some kind of a reduction of the cinematographic art to its most essential elements, a trip back to the origins of capturing motion and the invention of the first cameras.

Virtual spaces are mostly designed like a set of trade-offs and alliances between the sensuality of the virtual body, the presence of a virtual character, the presence of the viewer, and the spatial arrangement of machines producing images faster and in more seducing ways.

Designing virtual spaces and making movies can share similar goals, with each one following a different process leading to these goals:

1 Finding new ways to distribute camera locations in space

2 Focusing the viewer's attention on significant details

3 Walking through terrains held together by a story and creating connections, bridges between the ring enclosing the spaces of a story and the rest of the world

4 Moving forward in all possible directions inside a multidimensional world

Magicians use psychology to distract the audience from the trick being performed, and they use technological skills to perform their tricks. David Copperfield's mastery of technology allows him to execute parts of a trick that remain unseen by the audience. The suspension of disbelief required by the viewer during a magic trick is short compared to the duration of a movie. Putting together a puzzle of psychology and technology has always been a central element of fun fair attractions and circuses, from designing a roller coaster to displaying optical illusions. Movies are based on a similar illusion of perception. Moviegoers are under the impression that they view movement instead of single picture frames. Moviegoers perceive movement as the result of an alliance between a psychological factor, believing in an optical illusion, and a technological factor, making that illusion permanent over time.

CHAPTER 2
The Basic 3D Kit

1   THE ROADMAP

This chapter is an introduction to basic 3D modeling and animation in Maya and Virtools. "Basic 3D Kit" means that you will learn a selection of creative tools that will get you up and running quickly using Maya. This chapter, designed to be fun and creative, will hopefully make you anxious to explore this amazing software.

The following tutorials, created in collaboration with Miro Kirov, show how to use Maya and Virtools for creating interactive models and animated characters. We hope that you will enjoy the process of producing 3D interactive content.

The first tutorial is a step-by-step introduction to modeling and animating Mr. Cyclop, an interactive character in Maya. The second tutorial shows how to create trees and grass in Maya using 3D paint tools and how to set up Mr. Cyclop inside the forest in Virtools. The last part of the tutorial shows how to add scripts to the character in Virtools.

2   THE MAKING OF MR. CYCLOP, AN INTERACTIVE CHARACTER IN MAYA

Mr. Cyclop is a one-eyed mythological creature with a good and charming temperament. Mr. Cyclop is a goofy creature who enjoys playing inside a lush forest.

This tutorial covers step-by-step modeling, texturing, and animation of an interactive character, using two 3D applications: Maya and Virtools.

Maya will be used to build and animate the 3D content. Virtools will be used to add behaviors, including scripting an interactive environment using a keyboard, a mouse, or a game controller.

This tutorial shows how to create a goofy Mr. Cyclop using Subdivision Primitives, how to add textures, and how to animate the Cyclop's walk using skeletal structure and Inverse Kinematics systems.

2.1   Tour of Maya's Interface

Let's take a tour of the Maya interface.

- Open the Maya application, and take a tour of the interface.

- Maya, like other 3D applications, uses several modules called Modeling, Animation, Dynamics, and Rendering.

- From top to bottom, you can find the Menu bar, the Status Line, and the Shelf.

- The Workspace includes top, side, front, and perspective views.

- The mini Tools bar is located on the left side of the Workspace.

- The Channel box is on the right side of the Workspace.

- The Layers box is under the Channel box.

- In the lower part of the screen, you can find the Time Line, the Time Slider, and the Command Line.

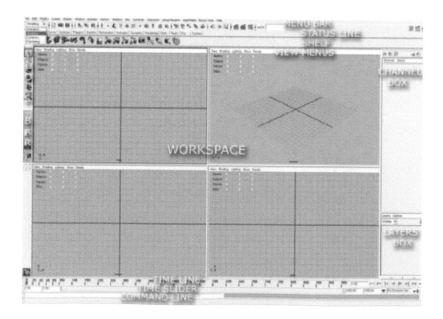

Maya's interface.

Let's review the row of tools represented with icons located below the top menu.

The tools change according to the module selected in the Status Line box. Click on the box located on the left in the Status Line to see the different modules. We will use Modeling and Animation for this chapter.

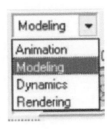

The tools change according to the module selected in the Status Line box.

The first group of tools found in the menu bar includes buttons for the following commands: Create New Scene, Open a Scene, and Save the Scene.

The second group of items includes buttons for selection: Selection by Hierarchy, Selection by Object, and Selection by Components.

The third group of items includes buttons for selecting different components: Vertices, Lines, Polygons, and Curves.

The fourth group of items includes buttons for snapping tools: Snap to a Grid, Snap to a Curve, and Snap to a Surface.

The last group of items includes shortcut tools for rendering: Render, IPR Render, and Render Globals.

The Shelf can be customized to store your most frequently used tools.

To customize the shelf:

Click on a tool in the menu bar and press Shift and Control simultaneously. You will see an icon of the selected tool added to the Shelf.

The Workspace is the place to preview your 3D content. You can customize the Workspace according to the type of views and rendering mode including shading, texturing, or wireframe mode. You can toggle between the perspective view and the four (top, front, side, perspective) views by pressing on the space bar. Each of the views has its own menu bar including View, Shading, Lighting, Show, and Panels.

To navigate in the perspective view window use the following commands:

* To rotate, use the alt key and the left mouse button.

* To pan, use the alt key and the middle mouse button.

* To zoom, use the right mouse button.

The Tool bar, located on the left side of the Workspace, includes the following tools, starting from top to bottom: Select, Lasso Select, Move,

Rotate, Scale, Manipulator, and Current, which displays the tool that you are currently using. The Layout bar with icons for different views of the Workspace can be found under the Tools section.

The Channel box, located on the right side of the Workspace, contains information about 3D models selected in the Workspace and current transformations applied to the models. The Layer box contains information about the layers created for 3D objects and lets you work on each 3D object without affecting the others.

The Time Line displays the key frame numbers. The Animation Controls and the Time Slider help you browse through the Time Line.

The Command Line for Maya's scripting language (MEL) commands lets you script transformations of the 3D content.

The Help Line is located below the Command Line.

After exploring the basic elements of the Maya's Interface, you can start building your character. You will use a fast and accurate way to model a character called Subdivision Surfaces. This modeling technique allows the computer to calculate shapes as polygonal surfaces and to display a subdivided smooth approximation of surfaces. After the modeling phase, we will convert the model to a polygonal mesh. Let's start modeling.

2.2   Modeling

Start your project by creating a project folder and its subfolders, where your content will be saved.

To generate a new project folder:

Go to the top menu, select, File > Project > New. In the dialog window specify a name for your project, choose the location where it will be saved, and click on Use Defaults. Click Accept to close the window.

To start creating the character, choose the modeling module. Go to the top menu, and select Create > Subdiv Primitives > Sphere.

Subdivision Surfaces has two modes: the PolygonProxy Mode, in which the shape can be manipulated as a polygonal mesh, and the Standard Mode, used for fine details.

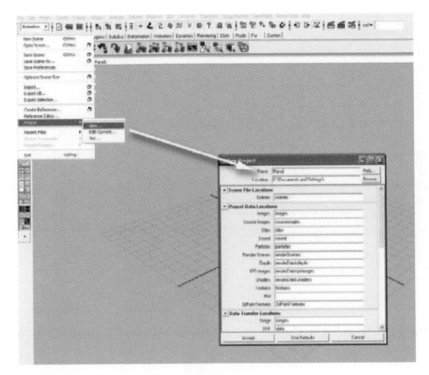

Start your project by creating a project folder and its subfolders, where your content will be saved.

We will use the PolygonProxy Mode for this tutorial.

Go to the top menu, and select Subdiv Surfaces > PolygonProxy Mode. The workspace displays a wireframe cube with its smooth approximation.

Press key 5 to display the sphere in the shaded mode.

Press key 3 to increase its smoothness.

Now let's select the cube by its components: vertex, line, and polygon.

In the Status Line click on the Select by Components button, then click on the Face button. The color of the cube changes to blue and a handle appears on each of the sides. Lines of bounding boxes are visible on the borders of the polygons.

To select the left side of the cube, click on the blue handle located on the left side of the cube. Press the Delete key. The 3D object looks like one-half of a sphere.

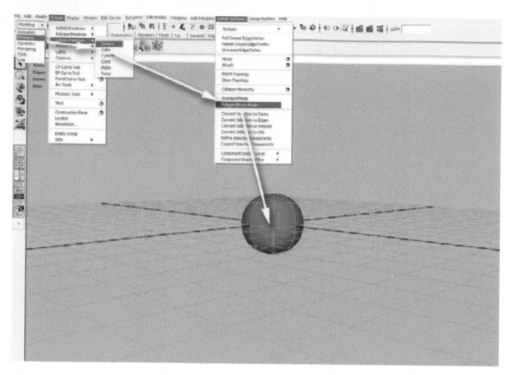

To start modeling, choose the modeling module. Go to Create > Subdiv Primitives > Sphere. We will use the PolygonProxy Mode for this tutorial.

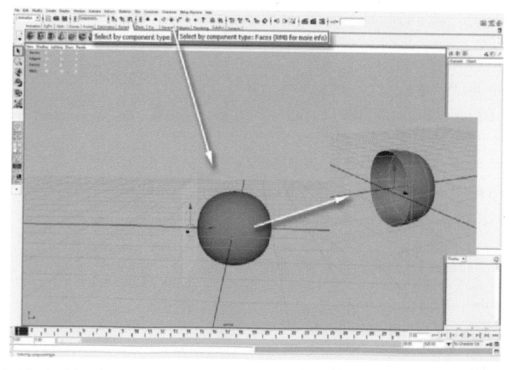

To select the left side of the cube, click on the blue handle located on the left side of the cube. Press the Delete key. The cube now looks like one half of a sphere.

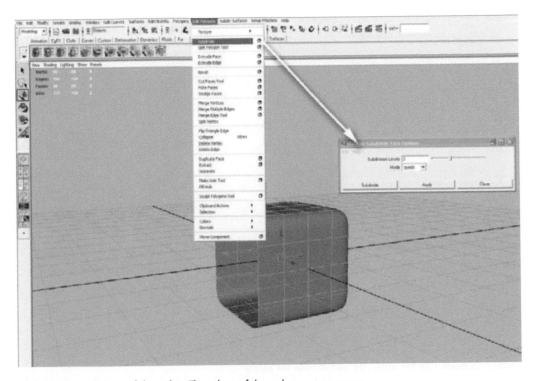

You can subdivide the polygons of the cube. The edges of the polygons turn green.

You can subdivide the polygons of the cube by selecting the Object Mode from the Status Line or by pressing the F8 key. The edges of the polygons turn green.

Go to Edit Polygons > Subdivide > Options. The Polygon Subdivide Face Options dialog window contains options for the Subdivide tools. Select subdivision level = 2 and click on Subdivide. Your model is similar to the one in the following illustration. First, you will be modeling only one side of the character. Later, you will mirror the other side of the 3D model.

This soft cube represents one half of the character's torso. You can use Move and Scale from the Tool bar to reshape the torso.

Go to the top menu bar, select, Polygon > Tools Options. Ensure Keep Faces Together is turned on.

In the Status Line, choose the Components Face mode. You can also press the F11 key. Select two polygons located on the right side.

Go to the top menu, and select Edit Polygon > Extrude Face.

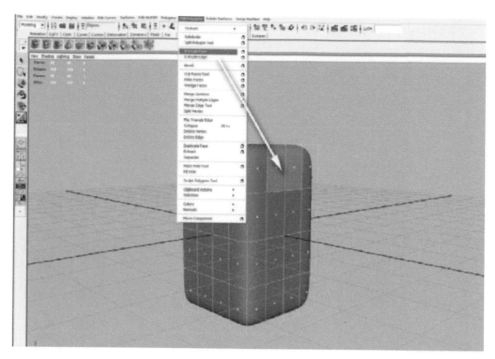

Select two polygons located on the right side and extrude the faces.

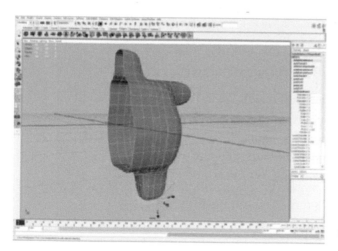

Click and drag the yellow arrow from the Manipulator tool, perpendicular to the selected polygons. You will notice that the smooth surface is changing shape locally.

You just created a new polygon that remains connected with the rest of the character. The Manipulator tool is automatically activated. Click and drag the yellow arrow, perpendicular to the selected polygons. You will notice that the smooth surface is changing shape locally. You began to create a new arm for your Cyclop.

Repeat the same operations for the left leg and the neck.

Let's play with selecting other components such as lines and vertices. You can select lines and vertices either from the icons of the Status Line or by pressing the F10 or F11 keys.

Let's model the character's limbs and face from the basic shape that you created.

Let's work inside the model to define the mouth and the eye socket.

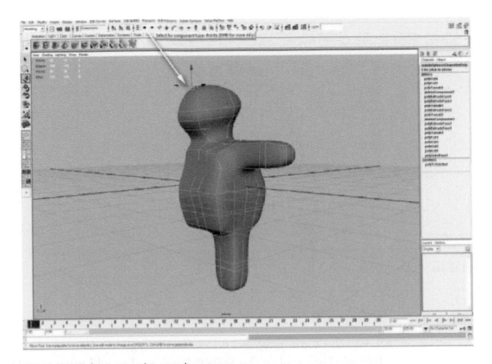

You can model the character by moving lines and vertices.

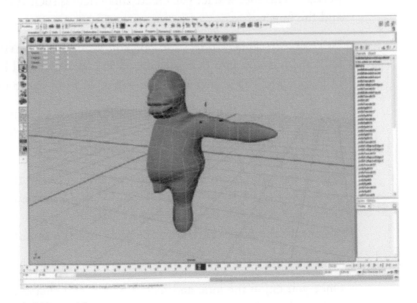

Modeling the character's limbs and face.

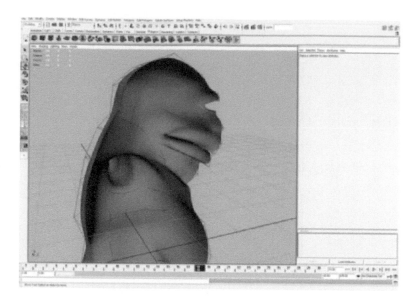

Side view of half of the character.

Go to the top menu, and select Edit Polygons > Extrude Face. Select the Extrude Face tool or the Split Polygon tool to add more details.

To split a polygon, select Edit Polygons > Split Polygon Tool. Select the Split Polygon tool, and click and drag on the desired lines surrounding the polygon. This tool will let you create small details such as the eyes, the nose, and the fingers.

To create the fingers, select the polygon located on the edge of the hand. Split the polygon into five polygons in the area that will become the base of the fingers. Three fingers will be extruded from five polygons.

To finish the detail of the fingertip, go to the top menu and select Edit Polygons > Poke Faces. Move the vertex located in the middle of the polygon at the fingertip to add a nice finishing touch to the fingertips.

Repeat the same operation to create the thumb. Select the polygon on the side of the hand. Use the Extrude Face tool, move the new polygon, extrude again, and poke the face. Move the vertex located in the middle of the polygon at the fingertip to get a nice finish of the thumb.

The same operations can be repeated for the leg and the toes.

Once you enjoy the look and feel of half of the character's body, you can convert the 3D model to a polygonal mesh.

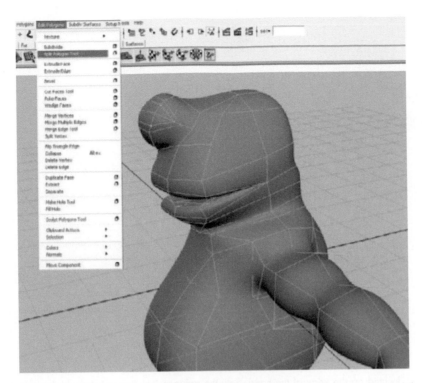

The split polygons tool will let you create small details such as the eyes, the nose, and the fingers.

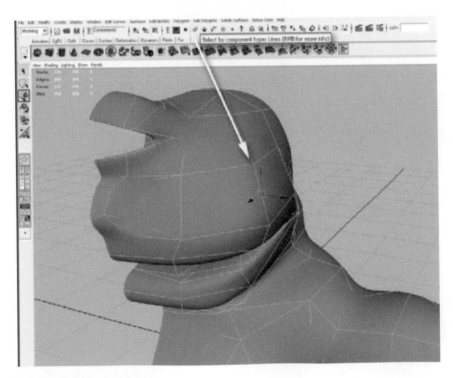

Selecting the edge of a polygon.

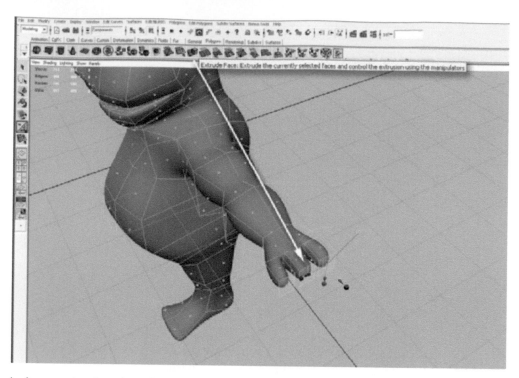

To create the fingers, select the polygon located on the edge of the hand. Split the polygon into five polygons in the area that will become the base of the fingers. Three fingers will be extruded from five polygons. Move the vertex located in the middle of the polygon at the fingertip to add a nice finishing touch to the fingertips.

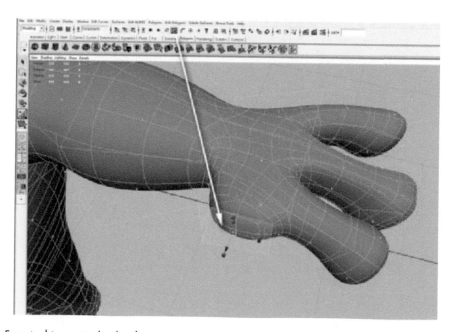

Use the Extrude Face tool to create the thumb.

To convert to a polygonal mesh:

Go to the top menu bar, select Modify > Convert, and check on the Subdiv to Polygons option box. The box is on the right of the name of the tool. A dialog window will open. Select the Tessellation Method with Uniform.

The Adaptive Method option may increase the tessellation inside the detailed area, which would create too many polygons. We will choose the Uniform Method because we want to keep down the number of polygons. Enter the following settings: Level = 1, the Division Per Face = 1, Replace Original Object = ON.

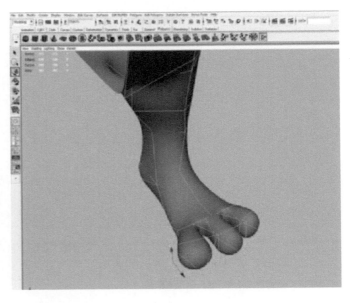

Creating the toes.

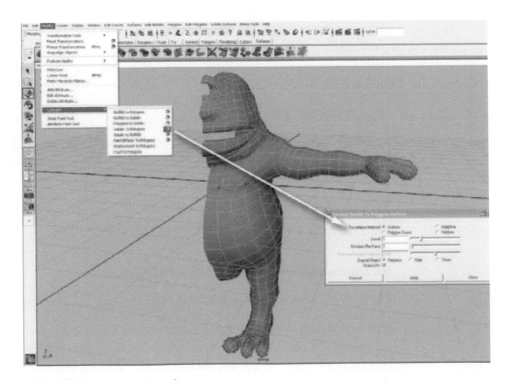

Converting the 3D model to a polygonal mesh.

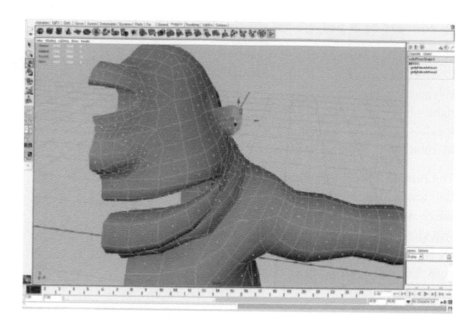

Let's add more details to the ears.

After conversion, you should have a shape made of interconnected polygons similar to the one you had before. Let's repeat the same modeling steps to add on more details for ears, nostrils, and wrinkles on the polygonal mesh.

Let's model the teeth.

Select the edges on the lips polygon located inside the mouth. Go to the top menu, select, Edit Polygons > Extrude Edge. Move the newly created edges.

Because your modeling job is detailed enough, you are going to duplicate the other half of the body to create a whole character.

To create the other half of the character:

Go to the top menu, select Edit > Duplicate Options.

In the dialog window, go to the column for the X axis, and change the scale value = −1. Click the Duplicate button. A complete mirrored clone is created. The edges of the two halves should slightly touch each other. If this is not the case, use the move tool to adjust one half.

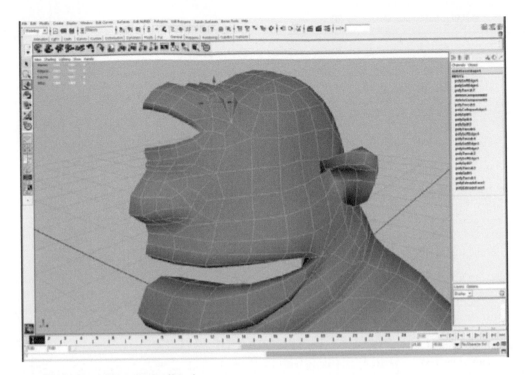

More wrinkles are added to the polygonal mesh.

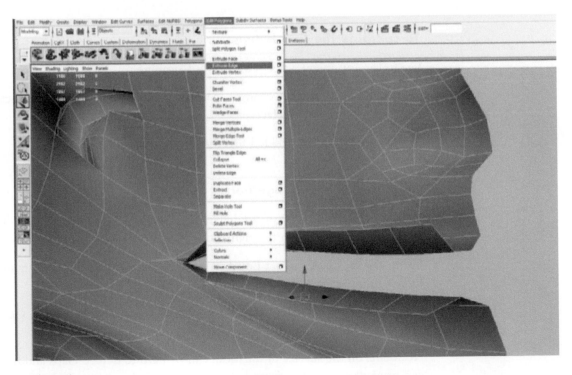

Creating the teeth.

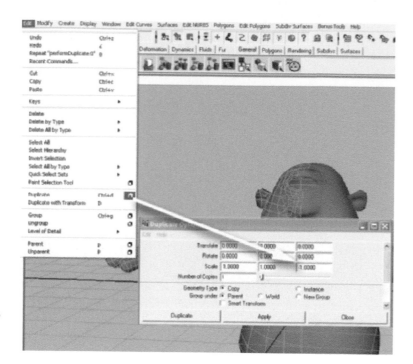

A complete mirrored clone is created. The edges of the two halves should slightly touch each other.

To combine the two halves:

Go to the top menu, select Polygons > Combine.

The color of the meshes changes from green and white to green, indicating that we have one surface.

The edges need to be welded together.

To weld the edges together:

Select the edges of each of the halves by clicking on one edge on the left half and on one edge on the right side.

Go to the top menu, select Edit Polygons > Selection > Select Contiguous Edges. Both edges will have a brown color.

The next step is merging two surfaces into one continuous surface. Go to the top menu, select Edit Polygons > Merge Multiple Edges.

Accidents on the surface of the mesh can be hard to detect. Gaps and holes can prevent some edges from being welded together.

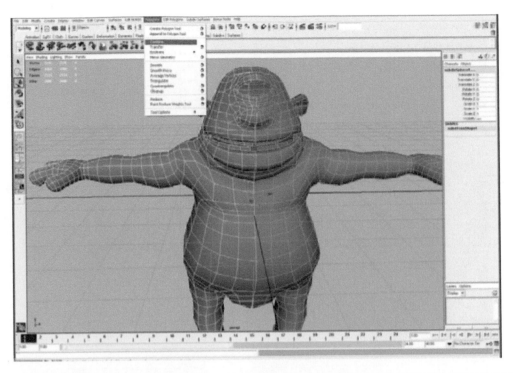

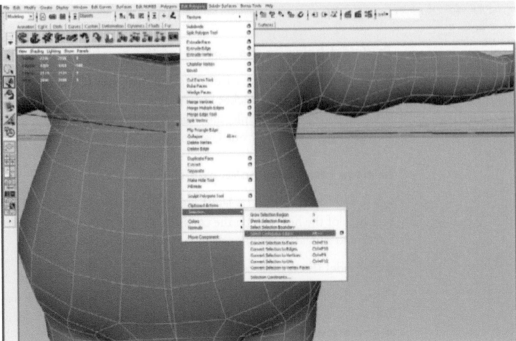

Contiguous edges are selected and change color.

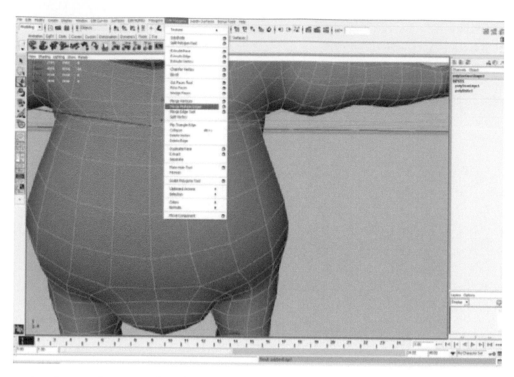

After merging the edges together, two surfaces become one continuous surface.

To detect holes on the surface of the mesh:

Go to the top menu and select Display > Polygon Components > Border Edges. Maya displays a thicker line around the contours of holes appearing in the mesh. In this example, the edges surrounding the mouth and the eye are obviously needed for our Cyclop character.

To close unwanted holes and gaps:

Go to the top menu and select Edit Polygon > Merge Edge. Click on the first edge of the hole and on the corresponding edge on the other side. Press Enter, and repeat the same process until the surface is closed.

Let's model the eye.

Go to the top menu, select Create > Polygon Primitives > Sphere. In the dialog window, specify the number of subdivisions along the height and around the Y axis = 10. Select Create.

Save your scene. Go to top menu, and select File > Save Scene As.

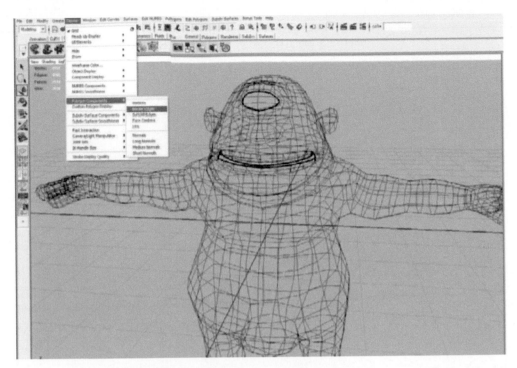

Maya displays a thicker line around the contours of holes appearing in the mesh. In this example, the edges surrounding the mouth and the eye are needed for our Cyclop character.

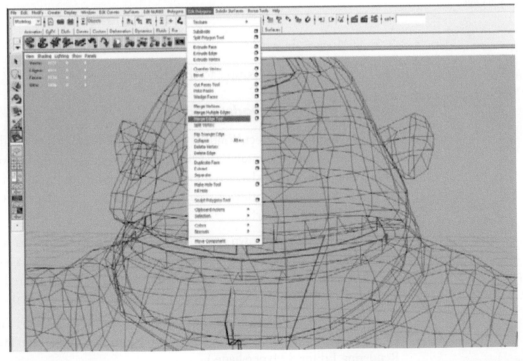

To close gaps and holes, click on the first edge of the hole and on the corresponding edge on the other side. Press Enter, and repeat the same process until the surface is closed.

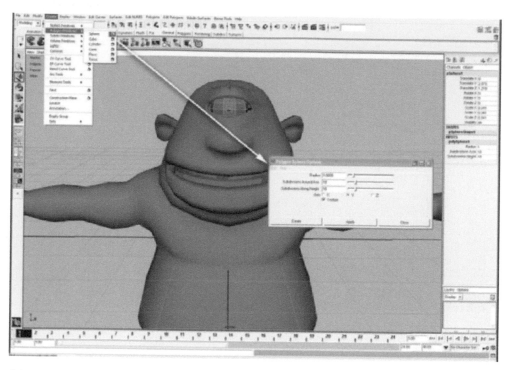

Modeling of the eye.

2.3   Materials and Textures

Let's apply colors and textures to the character.

To open the Hypershade window:

Go to the top menu, select Window > Rendering Editor > Hypershade. The Hypershade window contains the library of materials, textures, and utilities. 3D objects with surfaces require the creation of a material to receive a texture. When a new surface is created, Maya assigns a gray Lambert material by default to the surface.

You can customize your own materials and assign them to various surfaces of your project. In the case of this tutorial, we are using only image-based materials and textures. Maya can also use shaders, which are materials using code to describe the way the surface of the 3D model is rendered.

To create a new material for the eye, open the Hypershade window. (To open the Hypershade window, go to the top menu, select Window > Rendering Editor > Hypershade.)

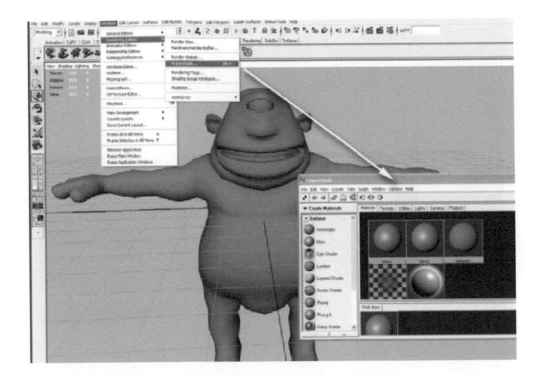

Go to the menu with materials and texturing tools found on the left side of the Hypershade window. Select Surface > Blinn. This material has a shinier surface, which is more appropriate for the skin than Lambert. The new material will be added to the library under the Material Tab. Double click on the Blinn material to open the Attribute Editor.

The Attribute Editor lets you edit most of the information and attributes of any 3D object in the scene including shapes, cameras, lights, materials, and shaders. In this case, the Attribute Editor displays Color, Transparency, Bump Map, and other parameters for the material. The number of folders and options sometimes can be overwhelming, but we will use only a small number of parameters in these tutorials.

Let's start with adjusting the color for the character's eye.

To get the white color of the eye material, move the Color slider all the way to the right. Rename the material "eyeSG" after Blinn in the Attribute Editor.

You need to assign the material to the eye surface to update the color changes in the scene. To get a white sphere for the eye inside the

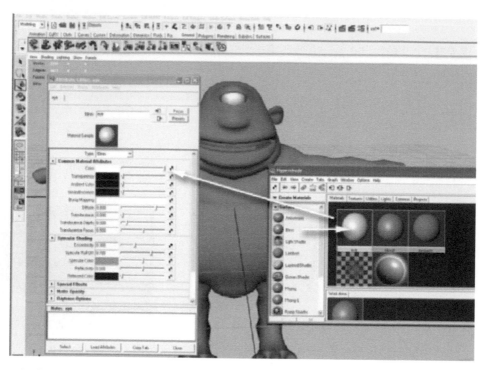

Adjusting the color for the character's eye.

Workspace, click with the middle mouse button over the material in the Hypershade window and drag it over the eye.

For Mr. Cyclop's body texture, go to Component Mode and select the polygons for the body; assign a Lambert shader from the Create Library. Repeat the same operation for the pants, and assign a new Lambert material.

For Mr. Cyclop's teeth texture, go to Component Mode and select the teeth polygons. This time you can choose a Blinn material. Press the middle mouse button and drag the material over the teeth.

Let's specify the UV mapping coordinates for every vertex of every polygonal face to apply textures to your model. UV mapping defines the corresponding coordinates of every vertex of the surface on a two-dimensional (2D) image. There are many ways to define the direction of UV mapping. You may use planar, cylindrical, or spherical mapping according to the shape of the object. You will use four textures for your character: one for the eye, one for the body, one for the pants, and one for the teeth.

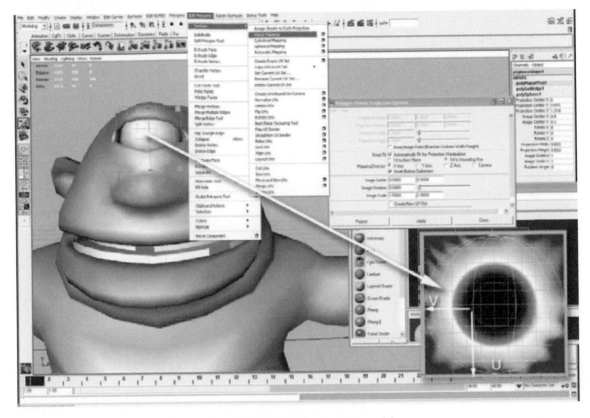

After initializing the UV mapping, the Projection Manipulator is displayed around the eye.

When creating textures, I suggest creating a 256 × 256 pixel image in Photoshop® and saving your work as a JPEG texture.

To initialize the UV mapping for the eye:

Go to the top menu, select Edit Polygons > Texture > Planar Mapping. Click on the check box located on the right of Planar Mapping. Using a texture with Planar Mapping on the surface of a 3D model has the same effect as projecting an image on a surface in front of the screen of a movie theater.

In the dialog window, check the options for Smart Fit and Fit to Bounding Box.

Select the perpendicular axis to the surface of the 3D model. In the case of the eye socket, the Z axis is selected. Click on Project, and the Projection Manipulator is displayed around the eye.

Previewing in the Workspace, several versions of the texture applied to the 3D model help you to find the best axis.

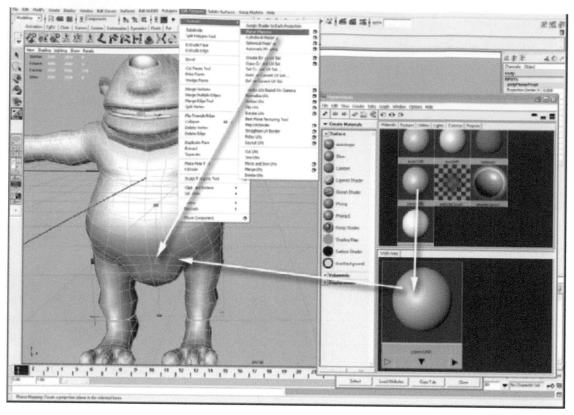

Defining UV mapping for the pants.

Let's define UV mapping for the pants. The type of projection is also Planar.

Now let's define UV mapping coordinates for the body.

The type of projection will be cylindrical. In the case of planar mapping, an image is projected in a movie theater on a round object placed in front of the screen. The image appears stretched on the sides of the round object but cannot reach the back of the object. In the case of cylindrical mapping, the image wraps around the round surface without deformation.

Although we are used to planar mapping, which is similar to the effect of a slide projector, cylindrical and spherical mapping cannot be recreated with a single projector in the real world.

Choosing textures for the surfaces of the character:

Go to Hypershader and double click on a material of your choice. A version of the Attribute Editor is displayed for each material.

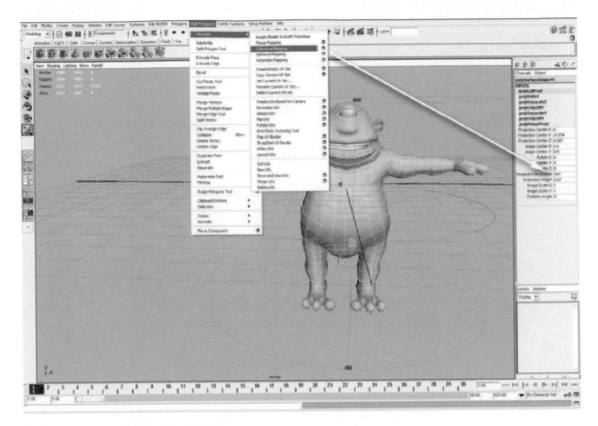

Defining UV mapping coordinates for the body.

On the right of the Attribute Editor, click on the check box icon on the right of the color slider. A new dialog window called Create Render Node opens. Click on the File button and upload an image of your choice by browsing through your files.

Viewing your texture in the Perspective window:

On top of the Workspace window select Shading > Hardware Texturing. Assign all the textures to the different parts of the character. For the teeth you also can apply texture and initialize the UV tool from the top menu. Go to Edit Polygon > Texture. This will display the texture on all faces of the selected polygons.

Important note: Elements in the scene may have different points of origin, which is the center of the world, described as X,Y,Z = 0,0,0 in your coordinate system. This may create problems when you export the scene to Virtools. The solution to this problem is to *assign the same point of origin for all 3D objects in the scene.*

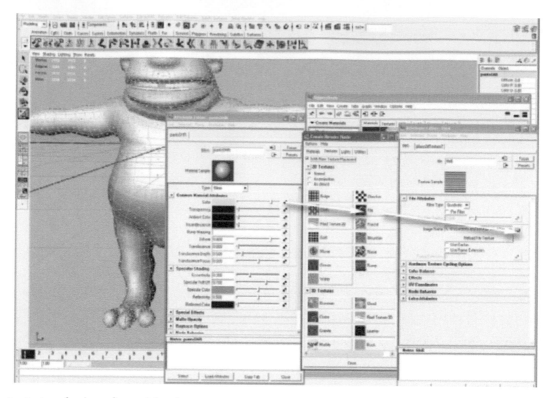

Choosing textures for the surfaces of the character.

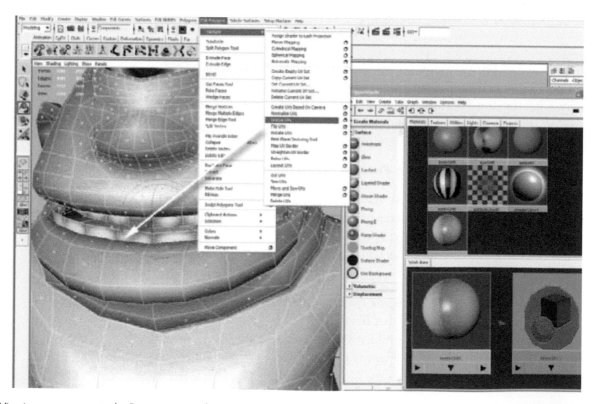

Viewing your texture in the Perspective window.

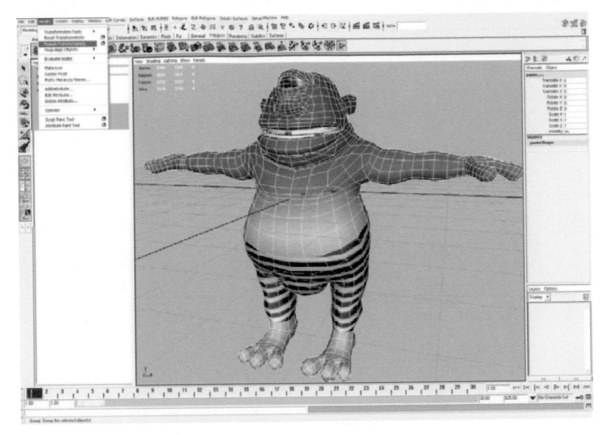

Assign the same point of origin for all 3D objects in the scene.

Select all the 3D objects by pressing Control + A keys.

Go to Modify > Freeze Transformations. The location coordinates for all the elements in the scene become X,Y,Z = 0,0,0.

Save your scene.

2.4   Building a Skeleton with Inverse Kinematics

It is time to bring your champion to life.

First, you will *build an internal skeletal system* that will control the body.

Select the Animation module on the right of the Status Line. Go to the Main Menu, select Skeleton > Joint Tool. Start drawing in the front view the spinal skeleton starting from the character's pelvis. To create bones, click once to create the first joint, then move your mouse to the location of the second joint and click a second time. A joint is created at every click. You

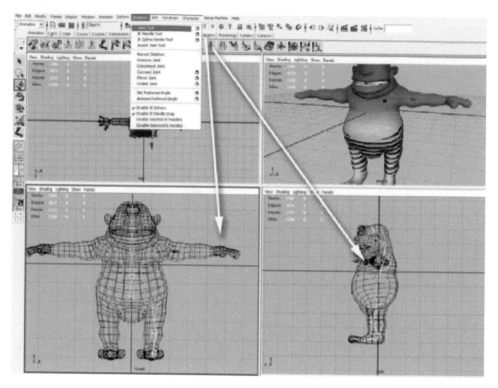

Building an internal skeletal system.

can create five bones from the spine to the tip of the head. Repeat the same for the right arm starting with the collar joint. Repeat the same for the right leg starting in the hip area. You can adjust the joints by clicking on them and moving them to the right location.

To connect the bones for the spine and for the leg:

First click on the hip joint then on the pelvic joint with the Shift key pressed. Go to the top menu, and select Skeleton > Connect Joint. Repeat the same to connect the arm-first-collar joint and the third spinal joint.

You can mirror the right arm joints to create the left arm joints.

To mirror joints:

Click on the collar joint, and then select Skeleton > Mirror Joint. Choose Mirror across YZ plane. Repeat the same operation for the right leg, starting with the hip joint.

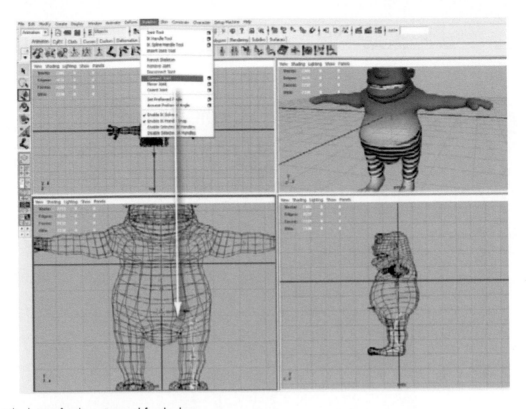

Connect the bones for the spine and for the leg.

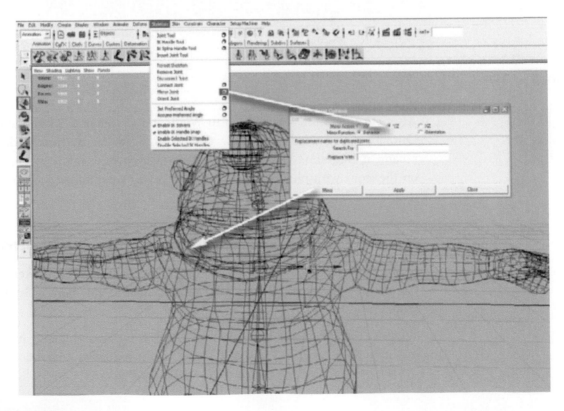

Mirroring joints.

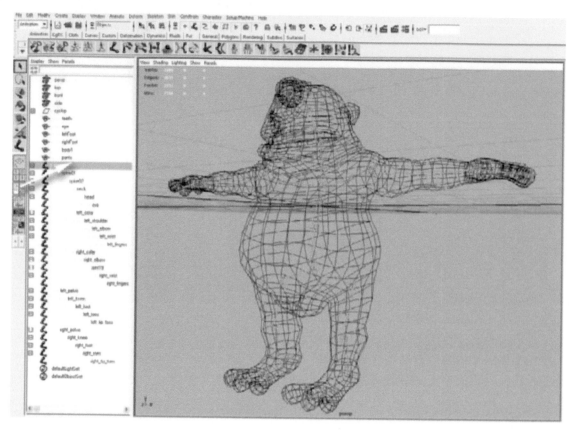

Renaming the skeleton.

After creating a full skeleton, go to Window > Outliner. Click inside the Outliner dialog window to view the nodes. Click on the first node to rename it "skeleton." Open the names for each chain of joints by clicking on the icons with plus signs next to the names of the nodes and rename them accordingly.

The next step is to bind the surfaces with the new skeleton:

Select all surfaces for the eye and the body and also select the skeleton. Go to the top menu and select Choose Skin > Bind Skin > Smooth Bind.

To test the binding process, select one of the joints and rotate it. The mesh should be stretched around the joint. In the following illustration, the shoulder area is stretched around the shoulder joint.

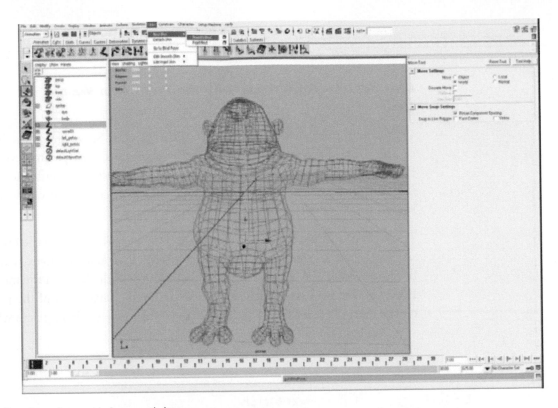

Binding the surfaces with the new skeleton.

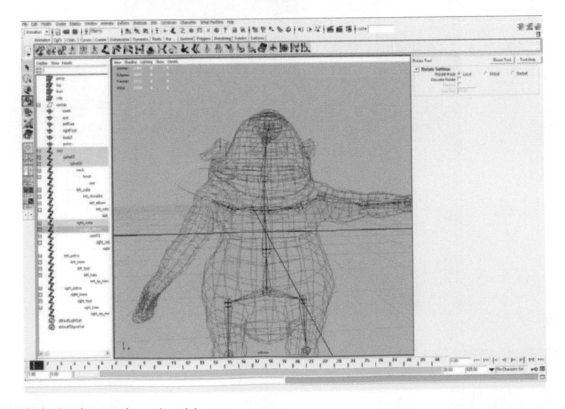

Testing the binding between the mesh and the joint.

Let's adjust weight maps with the Paint Skin Weight tool:

Go to the top menu and select Skin > Edit Smooth Skin. Check the box on the right of the menu item. Go to the dialog box with parameters for the joint, and click on different joints to see their respective influence on the skin of the character. Keep working only on the right half or the left half of the 3D model.

The color of the skin can change from white to black according to the intensity of the skin weight around the joint. Painting around the joint with a brighter color increases the value of the attraction of the joint on the mesh. Painting a brighter skin weight increases the force of attraction of the joint on the mesh. Parts of the mesh around the joint move easily in places with

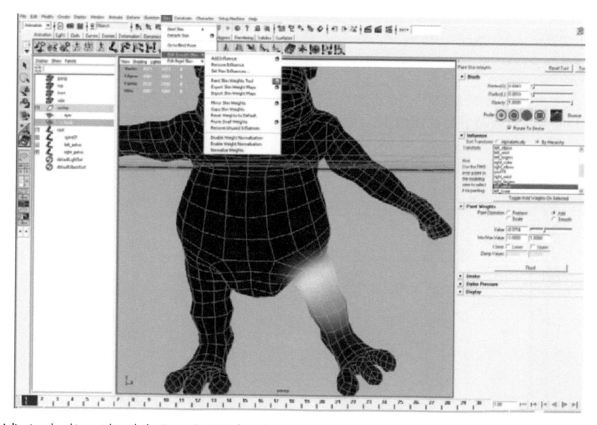

Adjusting the skin weight with the Paint Skin Weight tool.

a stronger force of attraction. Parts of the mesh located further away from the joint move with more resistance.

You can control the intensity of the attraction by changing the Min/Max Value from −1 to 1 and by turning on the Add button to paint over the skin.

Surfaces of body parts are not all affected by the motion of a joint. A skin weight with a dark color keeps meshes idle when a joint is moving.

For example, when you select the left collar joint, the skin weight for the head is black. This explains why the head will remain idle when you move the left collar joint. Select various joints in the Outliner and rotate them to test your model.

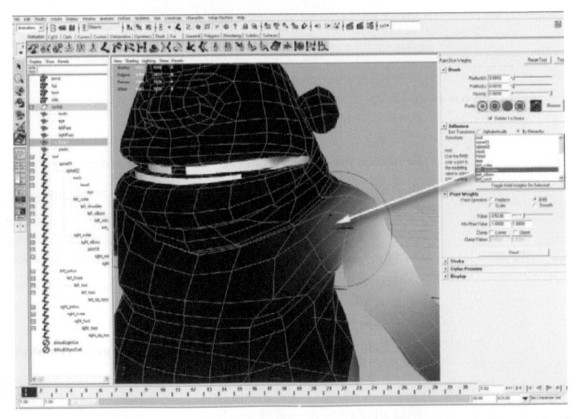

When you select the left collar joint, the skin weight for the head is black. This explains why the head will remain idle when you move the left collar joint.

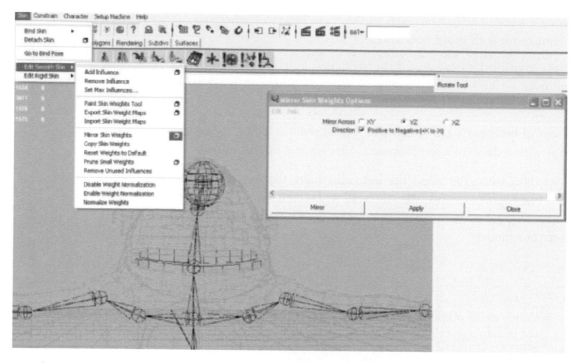

Mirroring the skin weights to the opposite side.

To mirror the skin weights to the opposite side:

Move your model in the middle of the Y axis. Go to the top menu.

First, select Skin > Bind Pose. Second, select Skin > Edit Smooth Skin > Mirror Skin Weights. Specify the right direction and a plane. If you work on the right side of the character, select Direction Positive to Negative.

In the workspace, run some tests by moving the joints with weight maps on the left and right sides. They should be identical.

Let's look at applying Inverse Kinematics to your skeleton:

Inverse Kinematics (IK) is an animation technique inspired from string puppets in which a chain of articulations follows the animation of a single string attached to the end of the chain.

A string, attached to the hand of a puppet, can connect a chain or joints including the lower arm, elbow, upper arm, and shoulder. Although only the last joint of the chain, the hand, is animated, the chain of joints follows the animation naturally. The IK technique allows you to animate only one joint located at the end of the chain of joints, instead of each joint of a skeleton.

You can use IK for animations controlled by real-time events. For example, one joint located in the palm of the hand can react to sounds produced by the viewer and move the arm in a realistic way.

To setup IK, go to the top menu, and select Skeleton > IK Handle.

First, click on the left shoulder joint. Second, click on the left wrist joint. Repeat the same operation for the right arm.

To set up IK for the legs, first, click on the left hip joint. Second, click on the ankle joint. Repeat the same for the right leg.

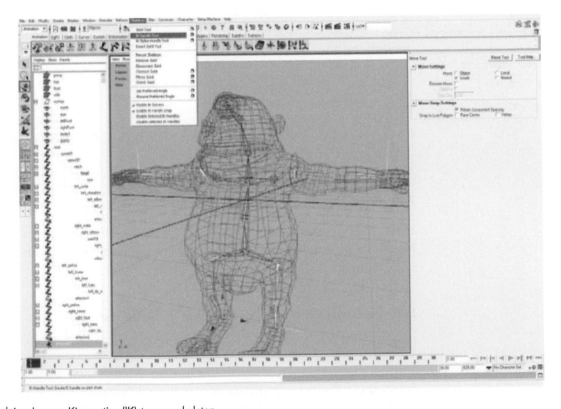

Applying Inverse Kinematics (IK) to your skeleton.

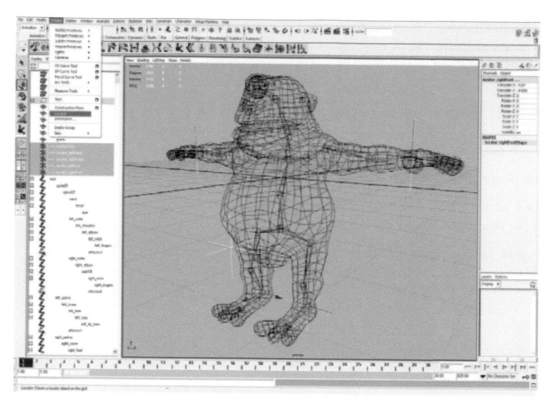

Creating locators.

Locators can help you add constraints to the IK handles when joints are difficult to select and move around.

To create locators:

Go to the top menu, select Create > Locators. Create five locators. Place four of them at the wrists and ankles. Place one locator for the body, in front of the model.

Let's constrain the locators to the IK handles:

First, select the locator. Second, select the IK handle.

Go to Constraint > Point. Select the box on the right side of the word. Select the following parameters in the Point Constraint Options dialog window: Weight = 1, Add Targets ON. Select Apply.

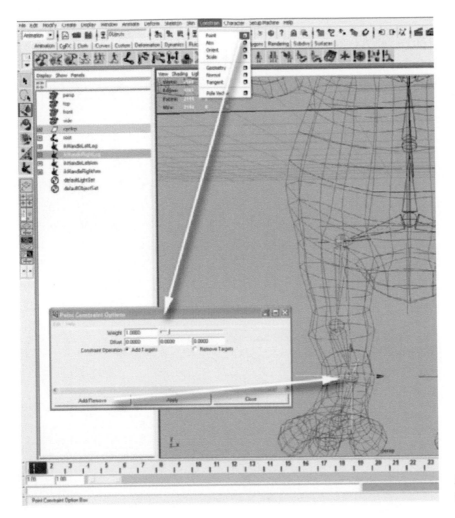

Setting up the locators on the IK handles.

Repeat the same steps for the locator in front of the body. Select the following parameters: Offset, Z = −3. Select Apply.

Before exporting the model to Virtools, you need to *reorganize the hierarchy of the nodes*.

In the Outliner dialog box, select all of the locators and group them by pressing the Control + G keys. Rename the group "Locators."

In the Outliner dialog box, drag the locators called Left Arm and Right Arm on top of the locator called Body.

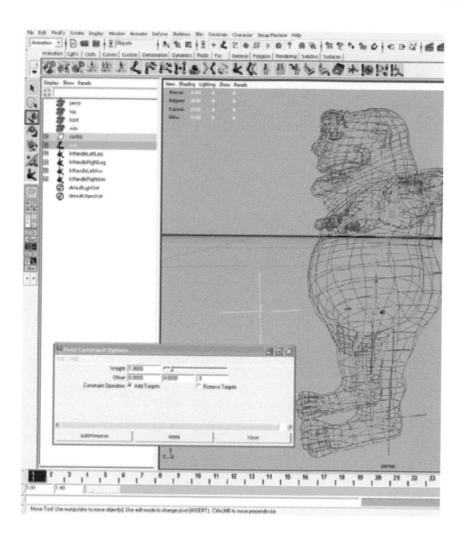

Setting up the locator for the body.

The following illustration shows how the hierarchy should be reorganized after moving the Body locator.

2.5   Creating a Walk Animation

Let's create a looped animation of a walk for the Cyclop character. First you create a walk cycle that you later loop over and over again to create the illusion of a continuous walk. This method allows you to use a short walk animation that will be optimized for 3D interactive content. The walk cycle takes 60 frames. The first and the last poses are the same so the cycle can be repeated without any visible discontinuity. The following

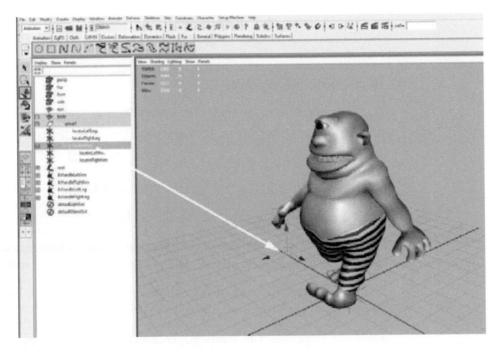

Reorganizing the hierarchy of the nodes.

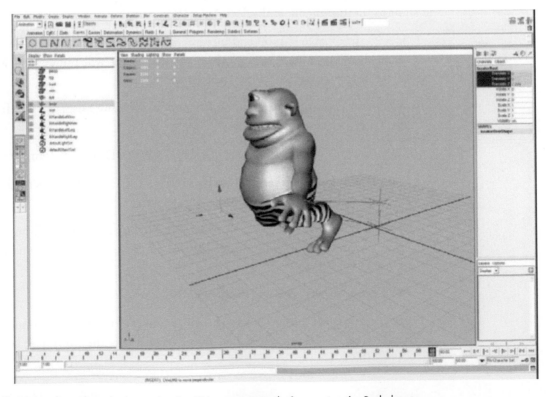

This illustration shows how the hierarchy should be reorganized after moving the Body locator.

illustration shows how to move the limb locators for the first posture of the Cyclop.

To animate Mr. Cyclop:

To key the first frame, first select the Body locator. Second, press the Shift + W keys. Repeat the same operations with the other locators:

Select the Left Leg locator. Press the Shift + W keys.

Select the Right Leg locator. Press the Shift + W keys.

The posture of the character is recorded at frame 1.

Select the Body locator, and move the character 15 units ahead. You can adjust the position of the leg locators to get a posture similar to the previous one.

You can specify the length of the animation in the End Time box located on the left of the Playback Animation Controls.

Move the Time Slider to frame 60, and record the posture by keying the locators again.

Go to the top menu, and select Animation > Animation Snapshot.

Specify time range = 1–60 and increment = 60. Press Snapshot. When you move the Time Slider back to the first frame you will notice that Maya took a snapshot of the character at frame 60.

Select the snapshot posture in the Perspective window. Display a wireframe version of the snapshot posture.

Go to the top menu, and select Display > Object Display > Template.

You can play back your animation by pressing the Play button located with the Playback Controls, in the lower right corner of your screen.

When you record the posture at frame 30, the legs and arms will be in the opposite direction of the postures previously created at frame 1 and 60.

Let's create two more postures:

One posture is at frame 15, and the other one is at frame 45. Fine tune the bending of the feet at these two frames, and press the Shift + E keys to record the keyframe.

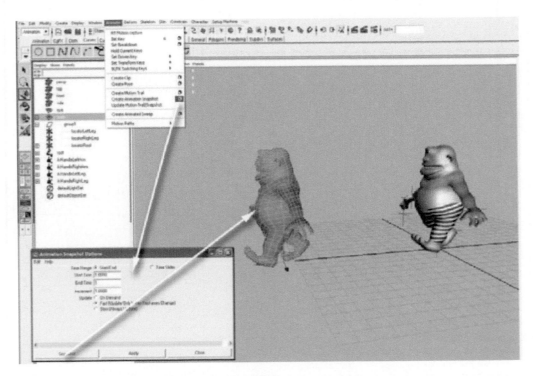

Select the Body locator, and move the character 15 units ahead.

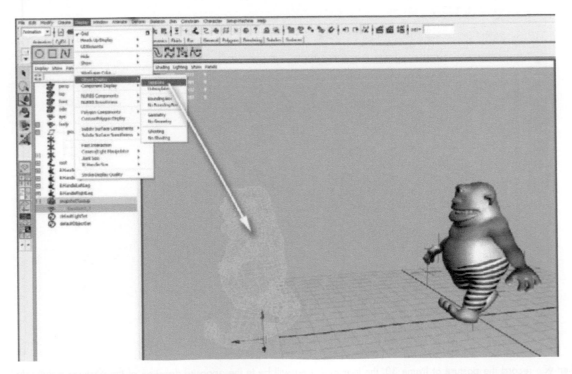

Display a wireframe version of the snapshot posture.

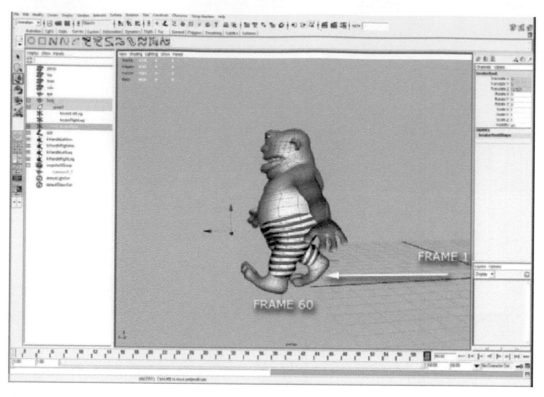

Play back your animation.

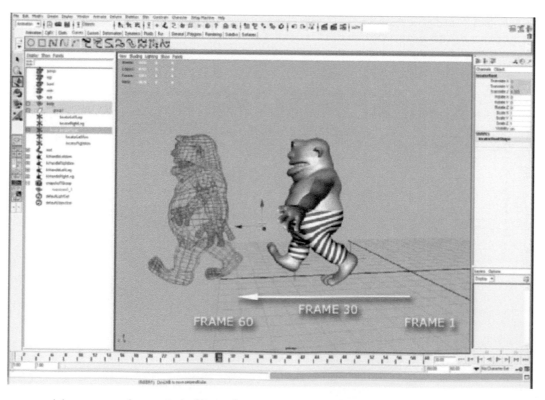

When you record the posture at frame 30, the legs and arms will be in the opposite direction of the postures previously created at frame 1 and 60.

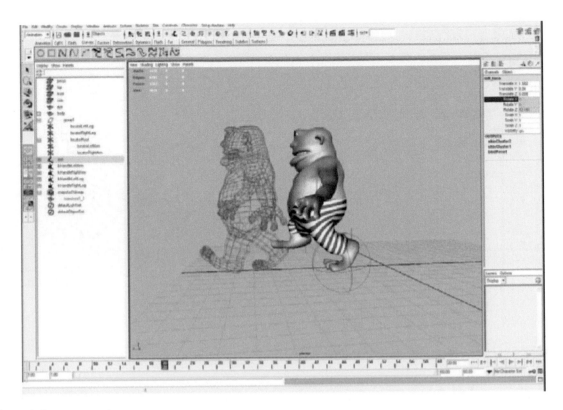

Let's create two more postures.

The Graph Editor helps you tweak the animation:

Go to the top menu, select Window > Animation Windows > Graph Editor.
Select the keyed locators, and in the Graph Editor choose View > Frame
All. The curve is smooth by default. You can make it linear by clicking on
the Linear Tangent button. Please remember to delete the snapshot posture
that you previously created.

To add a Point Light to your scene:

Go to the top menu, and select Create > Lights > Point Light. Save your work.

2.6   How to Export the Character to Virtools

Please make sure that the Virtools2Maya exporter plugin is installed on the
copy of Maya that you are using.

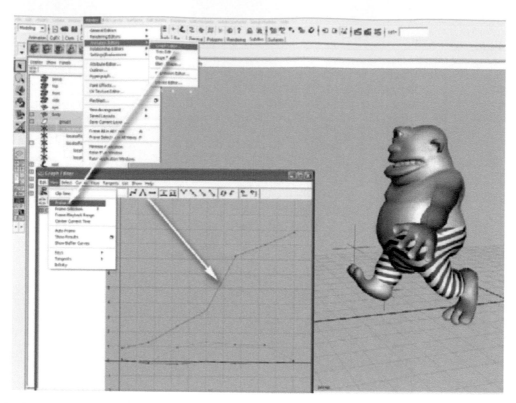

The Graph Editor helps you to tweak the animation.

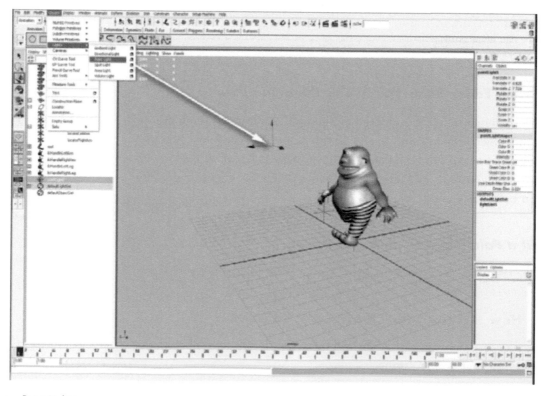

Adding a Point Light to your scene.

To activate the plugin, go to the top menu and select Window > Settings/Preferences > Plug-in Manager.

Select and enable the plugin.

Let's export the character to Virtools:

Go to the top menu, and select File > Export All. Select the box on the right side of the text. In the Export All Options dialog window, select the following parameters:

- Under File Type, select Virtools.

- Under Export Options choose

  - Type = Character,

  - Hierarchy = Full,

  - Export = All,

  - Exported Objects = Meshes, Lights, Normals, Textures, Cameras.

- Under Textures Options choose Include Textures in File = ON.

- Under Animation Options choose

  - Enable Animation = ON

  - Start = 0

  - End = 60 frames.

  - Sampling Step = 5

  - Frame Sampling = ON.

- Next to Animate, check Meshes, Lights, Vertices, Cameras.

- Name your file "exporting file" and export.

2.7 What Did You Learn in This Tutorial?

This section showed you how to create a fully functional character in Maya and to export the character from Maya to Virtools. The next section will take you step by step through the creation of a natural environment in Maya. You will be able to set up your character inside the environment and add interactivity in Virtools.

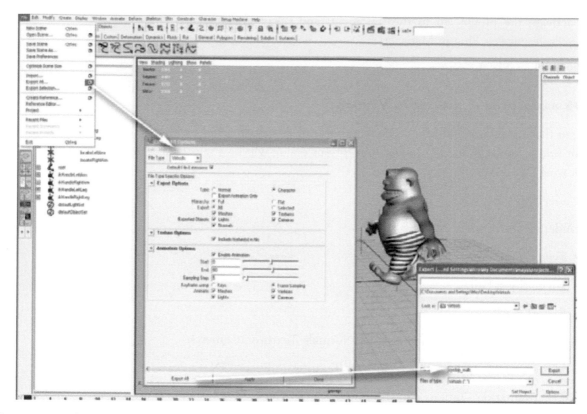

Exporting your character to Virtools.

3   CREATING A 3D IMMERSIVE ENVIRONMENT

This tutorial introduces you to Maya's 3D Paint Effects, allowing you to paint in 3D. A few strokes can paint trees, grass, or flowers. The paint strokes from your brush are converted into 3D objects inside a 3D space. Maya's ability to convert 3D Paint Effects to polygonal objects is helpful in creating content for interactive environments. This tutorial requires Maya 5.0 or higher.

You can control the strokes' shape, color, and density while painting on the horizontal grid plane of the perspective view or directly on other 3D objects.

Let's grab a brush and start to paint trees and grass in 3D. Let's create the ground for your forest.

3.1   Creating the Ground

- Open a new scene in Maya.

- Go to the top menu, and select File > New Scene.

- Select the Modeling module in the box located on the left of the Status Line.

- Create a polygonal plane for the floor of your scene.

- Go to Create > Polygon Primitives > Plane.

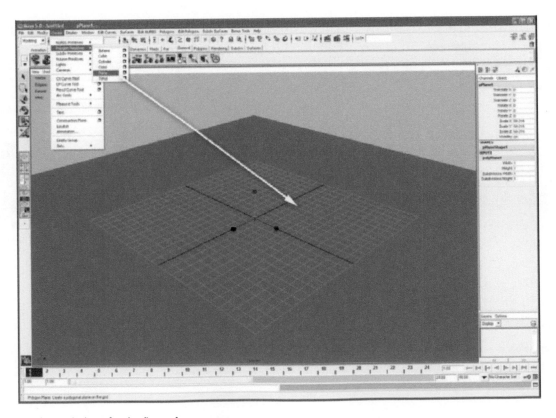

Create a polygonal plane for the floor of your scene.

3.2   Painting the Trees

A library of Paint Effects can be found in the Visor module.

To open the Visor window go to the top menu, and select Window > General Editor > Visor. Multiple folders with preset brushes are available. You can select a variety of paint effects inspired by natural forms by clicking on the folder named trees.

To resize your brush:

Go to the top menu and click on Paint Effects > Template Brush Settings. The Paint Effects Brush Settings dialog window opens. Select the Brush Profile. Set the Global Scale Slider = 2.0.

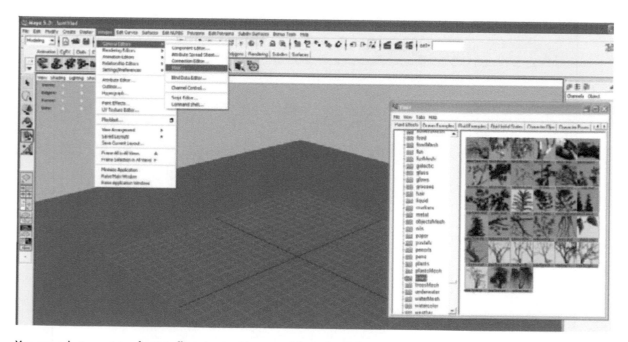

You can select a variety of paint effects inspired by natural forms by clicking on the folder named trees.

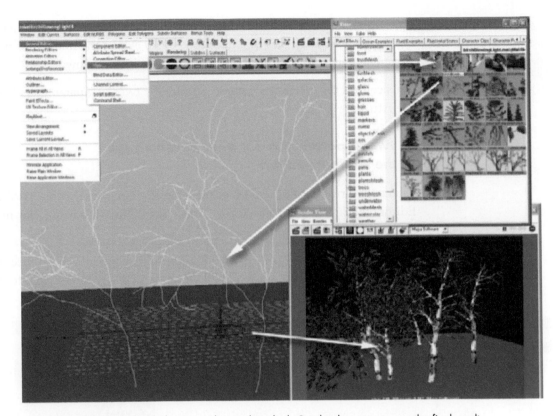

Maya creates randomly generated lush trees with a realistic look. Render the scene to see the final result.

Go to the Visor dialog window. Select Trees or MeshTrees. On the left side of the Visor window, click on the icon of the .mel file called birchBlowingLight.mel.

Go to the Workspace, and click and drag the mouse pointer a short distance near the Origin center in the Perspective window.

Maya creates randomly generated lush trees with a realistic look.

Render the scene to see the final result.

3.3　Painting Grass

Let's paint some 3D grass.

Go to the Visor dialog window. Select Grass on the left side of the Visor window, and click on the icon to choose the grass of your choice.

Important note: Because you will convert the content from your 3D painting into polygonal meshes, you have to keep an eye on the number of polygons. The density of the brush strokes needs to be kept low.

Let's control the creation of your vegetation by tweaking parameters inside the Attribute Editor.

To tweak parameters for your brush stroke in the Attribute Editor:

Select the brush stroke, and press the Control + A keys. In the Attribute Editor, check Sample Density and Tube Segments, which are the main parameters.

Two or three types of trees are needed for your scene in Maya. Once you import the trees in Virtools, you will clone them to create a whole forest.

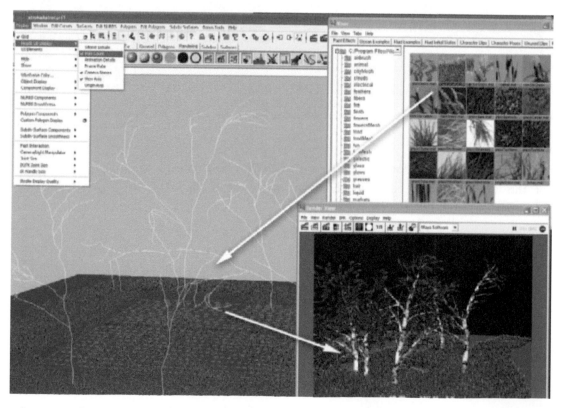

Two or three types of trees are needed for your scene in Maya. Once you import the trees in Virtools, you will clone them to create a whole forest.

The limit for the total number of polygons created inside a scene is 50,000. To view the polygonal count, go to the top menu, and select Display > Head Up Display > Poly Count. Check the number of Faces rendered in the scene.

3.4   Conversion of 3D Models to Polygonal Meshes

Let's convert the Paint Effects to 3D objects made of polygonal meshes.

To convert the 3D Paint Effect to polygonal meshes:

Select one of the strokes from the Outliner window or from the perspective view. Choose Modify > Convert > Paint Effects To Polygons. The output from the conversion is a high-resolution model with a high polygonal count.

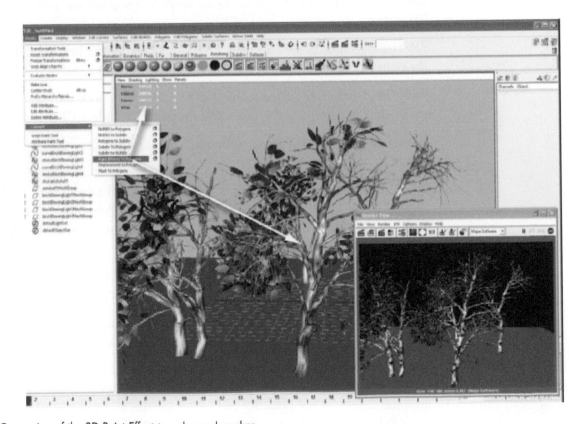

Conversion of the 3D Paint Effect to polygonal meshes.

3.5   Reducing the Number of Polygons

Let's see how Maya can help you reduce the polygonal count.

Select the Modeling module in the box located on the left of the Status Line.

To reduce the number of polygons for a tree:

Let's control the number of polygons for the tree.

Go to Display > Heads on Display > Polygon Count. Go to the Rendering menu and select Paint Effect > Paint Effects Mesh Quality.

Decrease the number of polygons by moving the sliders to the left for Tube Sections and Segment, and check the number of polygons on the Polygon Count. You will notice that this operation can radically alter the shape of the model. In most cases you will need to use the following polygon reduction to reduce the number of polygons while keeping the integrity of the 3D object.

Select a tree and choose Polygons > Reduce. Click on the box located on the right of the word to open the Polygon Reduce Options dialog window.

Enter the percentage of reduction. Go to Preserve. Check the following parameters, and keep Mesh Borders, UV borders, Hard Edges. Click Reduce.

Keep reducing the polygon count until your tree has less than 15,000 polygons.

If you get a message saying "Cannot Reduce Polygonal Object with Non-manifold Geometry," go to the top menu, and select Polygons > Cleanup. Click on the box located on the right of the word to open the Polygon Cleanup Options dialog window.

Go to Other, and check Nonmanifold Geometry and Normals and Geometry. Press Clean Up, and repeat the reduction process.

Please note that another way to lower the polygon count is to make quadrangular meshes. Maya creates triangular meshes by default.

You have created a 3D tree made of two 3D models, the branches and the leaves. Before exporting to Virtools, you should combine the two models into one. Select all the branches and leaves for the tree and go to Modeling > Polygons > Combine.

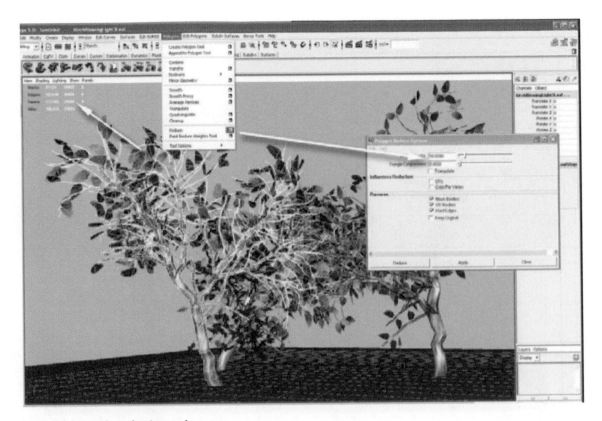

Reduction of the number of polygons for a tree.

3.6   Materials and Textures

Let's open the Hypershade window to assign materials to your 3D models.

To open the Hypershade window, go to the top menu, and select Window > Rendering Editor > Hypershade.

Name the leaves material "LeavesSG." Name the trunk material "TrunkSG."

To select one material in the Hypershade window:

Go to the main menu in Hypershade. Select Graph > Input and Output Connections. You will see a network of connections that Maya generated after converting the trees to polygons.

You are looking for more information about textures, so double click on the texture icon and check the path of the texture that is used for the trees.

Selecting one material in the Hypershade window.

The file is stored in Program Files > Maya > Brushes > Trees > pine2LeafShader.

The texture used for the leaves is called sideleaf.iff. For the bark of the tree the texture is called wrapBark.iff.

Save your scene as "forest."

3.7   How to Export the Scene to Virtools

Maya will export your polygonal mesh data, plus the UV mapping information and shaders, generated during the conversion process.

You need to export each element one by one as one tree or one object, .obj (object) files.

To export the scene to Virtools:

Select the object and go to File > Export Selection.

Point your file browser in the desired location (for example, the data folder located in the Virtools database). Assign a unique name, choose a file type (.obj), and click export.

Let's export the elements of the scene to Virtools:

• Select one of the trees and go to File > Export Selection.

• In the exporting window, select Virtools.

• Check Default File Extensions.

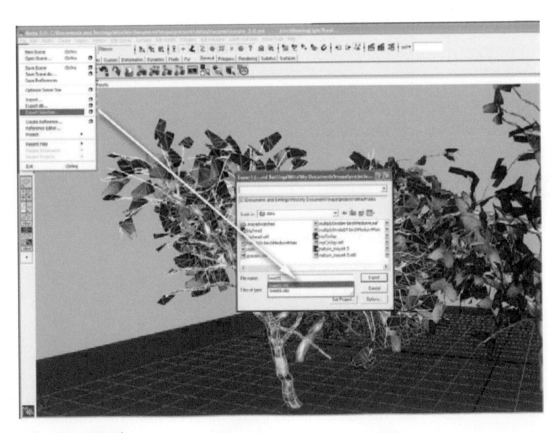

Exporting the scene to Virtools.

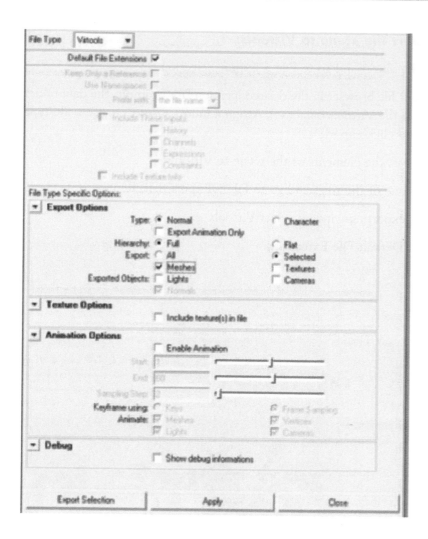

The Virtools export dialog window.

• Under Export Options select the following parameters:

    • Check Type = Normal,

    • Hierarchy = Full, Export = Selected,

    • Meshes and Normals are on.

3.8   Setting Up the Scene in Virtools

Open Virtools. The 3D Layout window with a perspective view is on the left. The Building Blocks folders are on the right. The Schematics window is at the bottom. This window displays a network of connections between building blocks that is similar to the Hypergraph window in Maya. The Lever Manager is similar to the Outliner window in Maya.

You will create a Data Resource folder with subfolders to save your scene.

To create a Data Resource folder:

Go to Virtools top menu, and select Resources > Create New Data Resource.

Name the Data Resource folder and save it. Virtools creates a file called Data_Ressource.rsc.

The files exported from Maya or from other 3D software applications can be saved in the subfolders inside the Data Resource folder. For example, a character will be saved in Data Resource folder > Character. A tree, which is a 3D object, will be saved in Data Resource folder > 3D Entity.

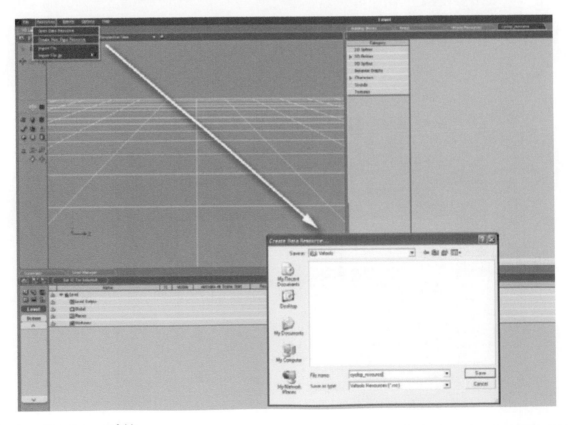

Creating a Data Resource folder.

To access your assets saved in the Data Resource folder in Virtools:

Go to the top menu, and select Resources > Open Data Resource. Select the Data_Resource.rsc file.

Please note that you can also import in Virtools, one by one, all the files that you exported from Maya in .nmo file formats.

To import .nmo files one by one:

Go to the top menu, and select Resources > Import File.

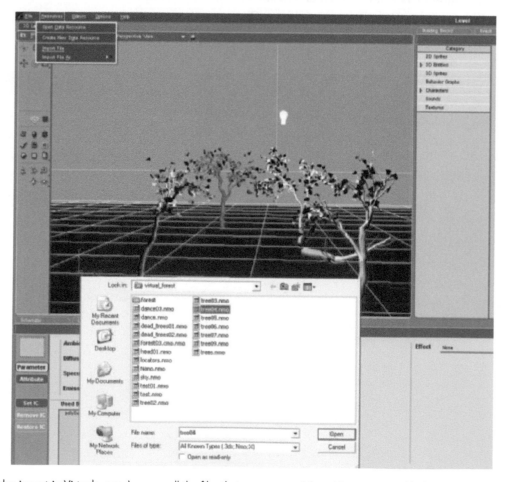

You can also import in Virtools, one by one, all the files that you exported from Maya in .nmo file formats.

3.9   Loading Textures from Maya in Virtools

In the Level Manager, go to Globals > Materials. Double click on the name of material. The Materials Setup window shows materials from 3D objects in the scene.

Although most of the time Virtools loads 3D models with their textures in place, some textures may not be displayed automatically, which is the case with 3D objects created in Maya.

The following steps will help you to correct this situation.

To export textures from Maya to Virtools:

Let's export textures from Maya.

Textures in .iff format need to be converted to .jpg format using the Maya FCheck converter.

To convert a texture from Maya:

Open the FCheck application from your desktop. Go to the Start Menu > Programs > AliasWavefront > Maya > Fcheck.

Go to the Fcheck menu, and select File > Open Image. Load the leaf and bark .iff files. Go to File > Save Image. Save as .jpg file formats. The texture files are saved in the Data Resource folder > Textures folder that you previously created.

In Virtools, select the tab for the Data Resource folder located on the right of the 3D Layout window. Select the Textures shelf. The names of the textures that you previously saved will appear in the column to the right.

To load a texture in Virtools:

Click on the name of a texture and drag it over the 3D Layout window or Level Manager window. The texture will automatically reside in the Level Manager window, under Level > Global > Textures shelf. When you save your file, assets referenced inside the Level Manager window will be saved with your project regardless of the presence of the Data Resource folder.

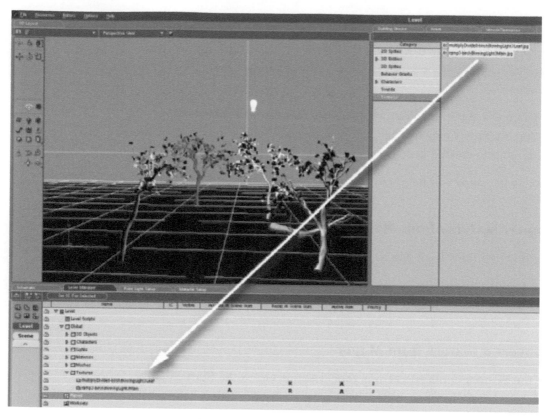

To load a texture in Virtools.

If you don't see the Resource folder located on the right of the 3D Layout window in Virtools, you may need to load the Resource folder. Go to the top menu, select Resources > Open Data Resource. Select the Data_Resource.rsc file.

Sometimes, textures are missing on 3D objects viewed in the 3D Layout window. If the texture is missing from Level Manager > Textures, you need to import the texture again. If the texture can be located in Level Manager > Textures, you need to associate the texture to the 3D model again.

To reassign a texture to a 3D object in Virtools:

Go to the 3D Layout window, and right mouse click on various shades visible on the trees. In the drop down menu, select Material Setup.

A Material Setup tab opens at the bottom of the 3D Layout window. The material is located in the lower part of the screen. Go to Texture, and select the appropriate texture for the material that you selected.

You are looking at trees inside a Virtools scene with the same look and feel of the trees that you created in Maya.

Assign a texture to a 3D object in Virtools.

3.10   Creating a Skybox

Let's create a skybox for your scene in Virtools. More in-depth examples of skyboxes can be found in the chapter covering textures.

To create a skybox:

Go to the top menu, and select Resources > Open Data Resource. Open the Virtools Resources folder installed with the application, which can be found on your hard drive at Program Files > Virtools > Documentation > VirtoolsResources.rsc.

Go to the right of the 3D Layout window. A new tab called VirtoolsResources appears next to your existing Data Resource. Check Virtools Resources > Textures > Sky Around Subfolder, and select the five textures in .tiff files format. Drag the textures over the 3D Layout window or Level Manager window.

In the Building Blocks tab, select World Environment Folder > Background Subfolder. Select the Sky Around building block, and drag it over one of the cameras located in the Level Manager > Cameras. Please note that this building block can be connected to any 3D object in the world.

In the Sky Around window select the corresponding textures: sky_t for the top, sky_b for the bottom, sky_d for the back, sky_f for the front, sky_l for the left side, and sky_r for the right side of your environment.

Run the program by clicking on the Play button located at the left corner of the Virtools window. Textures of the sky are projected around the scene.

3.11   Using Textures with Alpha-Channels

You may notice that the textures for the leaves are unrealistic. A whole polygon is showing around the leaf texture. You can fix this problem by adding an alpha-channel to the texture. This channel will turn black areas of the texture transparent.

To add an alpha-channel to your texture, go to the Level Manager and double click on the leaf texture. The leaf texture tab will open under the 3D Layout window.

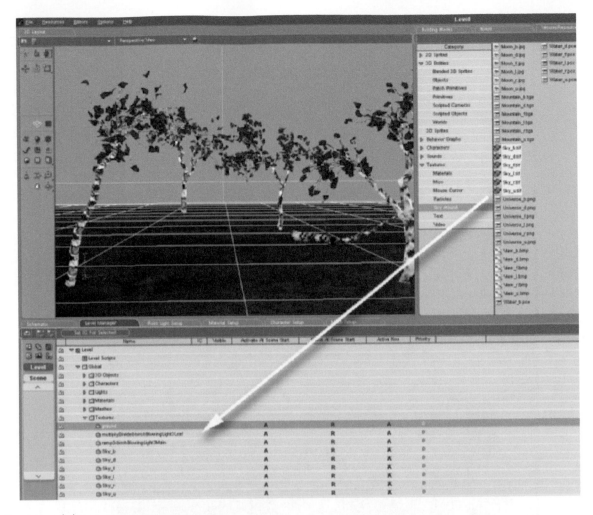

Creating a skybox.

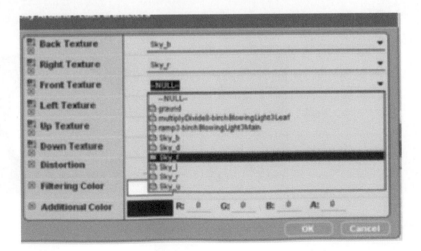

Selecting textures for the Sky Around building block.

Select the following parameters:

- Check Color Key Transparency.

- Use the color picker to select the black color inside the leaf texture.

- Check the transparencies in your scene.

Adding an alpha-channel to your texture helps to create leaves with transparencies.

3.12　Importing a Character

Let's import Mr. Cyclop.mno, the character created in the previous tutorial.

Go to the top menu, and select Resource > Import File As > Character. Go to the Level Manager. Check Characters > Cyclop subfolder, where you can find Mr. Cyclop's body parts, bones, and animations.

Please note that you need to export Cyclop as a character in Maya if you want to use Cyclop as a character in Virtools. If you can't get Virtools to recognize Cyclop as a character, go to Maya and repeat the steps from Section 2.6.

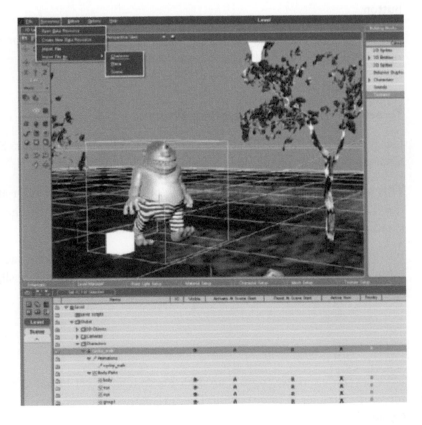

Let's import Mr. Cyclop.mno, the Maya character created in the previous tutorial.

3.13   The Character Stands on the Floor

Let's make our character stand on the floor. This may be critical when your character stands on an uneven surface, such as a hill. The character is going to receive a force of attraction similar to gravity. You need to prevent the character from falling across the ground.

Let's see how you can declare that the ground is a Floor, add gravity to the scene, and keep the character standing on the ground.

To add the Floor attribute to the 3D Object called Ground:

Go to the Level Manager > Level > Global > 3D Objects > Ground. Ground is what we called the plane created in Maya in Section 3.1 of this tutorial.

Double click on Ground. A new tab will open with the 3D Object setup. Click on the Attribute button. Click on the Add Attribute button.

The Add Attribute dialog box opens. Select Floor Manager > Floor. Click Add Attribute. Click Add Selected, and close the window.

To set up the character on the floor:

Go to Building Blocks, located on the right side of 3D Layout. Select Characters > Constraint Folder > Character Keep on Floor.

This building block looks for any obstacle or 3D Object with a floor attribute. The behavior will add gravity to the character and keep the character standing on the floor.

Please note that because the behavior is activated only during playback, the initial position of a character with Keep on Floor behavior should be slightly above ground.

3.14   The Character Walks

Let's get our character to walk around. Remember to set Initial Condition for your character before starting. The interactive animation setup for Mr. Cyclop has two steps:

Adding the Floor attribute to the 3D Object called Ground.

Creating an interactive walk animation for Mr. Cyclop:

Go to Building Blocks, located on the right side of 3D Layout. Select Characters > Movement > Character Controller.

A Character Controller dialog window opens. Select Walk Animation = cyclop_walk animation. Click OK.

Controlling the walk animation for Mr. Cyclop:

Go to Building Blocks, located on the right side of 3D Layout. Select Controllers > Keyboard > Keyboard Controller.

Drag the Building Block over the character in 3D Layout. The character is highlighted with a yellow bounding box. The Keyboard Controller Building Block uses the numeric keys 8 and 6 for playing animations and

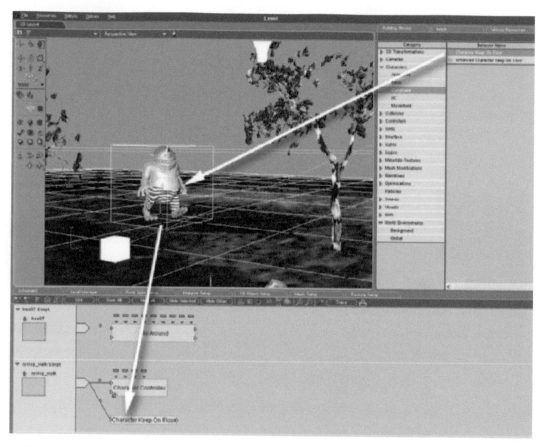

Setting up the character on the floor.

the numeric keys 2 and 4 for turning the character to the right or to the left.

In this example, press key 8 to trigger the walk animation.

Play the application, and check the walk animation by pressing the numeric key 8. Your Cyclop stays on the floor and walks. Press keys 2 and 4 to turn the Cyclop.

To save your creation as a cmo file, go to File > Save Composition.

To bring the application into the Virtools web player, you can publish a compressed .vmo file, embedded inside a web page.

Go to File > Create a Web Page. Choose the window size for your 3D inter-active content and select Preview in Browser.

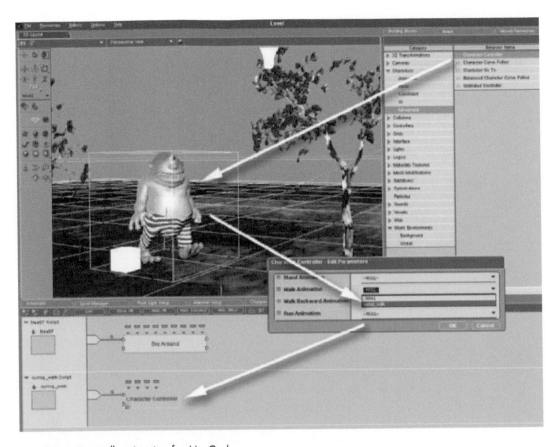

Creating an interactive walk animation for Mr. Cyclop.

Virtools generates a Virtools Web player file with .vmo extension, and a Web page with .html extension. The Web page contains the code to embed the Web player file.

So this is it! Congratulations on completing your interactive 3D application.

3.15 What Did You Learn in This Tutorial?

This tutorial showed you how to create a small forest in Maya and how to export 3D objects with their textures from Maya to Virtools. You learned how to set up an animated character inside a 3D interactive scene and how to get the character to walk around.

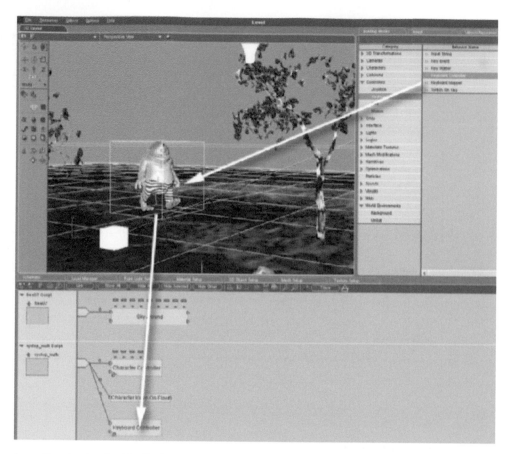

Controlling the walk animation for Mr. Cyclop.

This chapter gave you a roadmap of how to produce 3D interactive content. The next chapter about kinematics will show you how to create virtual worlds with elements similar to the ones from this basic 3D kit. You will find more possibilities for interaction between the character and the forest. You will animate the forest. Mr. Cyclop will find ways to avoid obstacles on his way and he will also play music with the trees.

CHAPTER 3

Interactive Textures and Lighting Design

1   INTRODUCTION

Textures play a critical role in the overall design of virtual spaces. Textures and lighting work together to bring emotion and dramatic tension to scenes. Interactive textures go beyond the use of 2D images carefully applied on the surface of a 3D object; they help to manage the viewer's dramatic experience inside a virtual space. They are often used as 2D media placed between the mesh of a 3D object and the lighting and environment of a virtual scene. This chapter covers various kinds of interactive textures presented in order of increasing complexity.

1.1   Bitmaps and Procedural Images

The first group of tutorials includes interactive textures using **still images** and **video textures** created from videos, animations, prerecorded movies, live video, or live images filmed by a virtual camera inside the virtual world.

Still images can be mapped on a 3D object like applying a decal on a toy. The placement of textures assigned to a 3D object follows a predefined texture map, which is a topography of surfaces (called materials in 3D Max and shaders in Maya) covering specific parts of a 3D object. For example, several textures applied on a 3D face will follow the arrangement of surfaces or materials created for the skin, lips, eyes, and teeth.

Movies made of sequences of images, usually displayed on a flat screen, can also be mapped onto the surface of a 3D object. Canned movies are created with prerecorded or prerendered videos or animations. Live movies are recorded on the fly with a camcorder or a Webcam. In both cases, movies can be textured the same way as still images regardless of the shape and geometry of the 3D model. We will look at examples of 3D objects textured with a live video coming from a camcorder or a Webcam. In other examples, the video is filmed by a virtual camera shooting images inside the virtual world. We will also cover interactive textures processing a video signal pixel by pixel. The video signal can be noise generated by the computer.

1.2   Shaders

The second group of tutorials includes **shaders** with textures made of bitmaps and procedural textures processed with code or created from scratch with code. Shaders can modify most of the parameters available for textures: ambient, diffuse, specularity, transparency, reflectivity, and emissivity. Shaders can radically change the relationship of an object with its environment by creating more complex behaviors where lighting, textures, and reflections interact together.

Shaders can change the perception and the mood of a scene by acting as a filter between the surface of a 3D object, the lighting, and other environmental effects. For example, a chrome reflective shader can turn a 3D object into a perfect mirror. Shaders can operate on vertices and pixels at the same time. The attributes for each vertex inside a 3D object include: 3D information about position, color, and 2D information about texture coordinates. Stencil-shadowed texture, cartoon-shadowed texture, chrome reflective models, or even models with fur are applications of advanced shaders covered in this chapter. We will look at interactive shaders that can modify the way pixels are rendered on the object's surface.

Interactive textures take advantage of the possibility of processing the images used for texture mapping. Procedural textures include coded procedures that address pixels inside a bitmap image or from an image generated by code. Procedural textures are made possible by the architecture of fully programmable rendering pipelines—for example the OpenGL pipeline. Interactive textures, created from 2D images, bitmaps, sequences of bitmaps, or procedural images created from code, can instantly change the shape, appearance, and motion of a 3D object.

Texture attributes can have different names depending on the software being used. The following list shows examples of words commonly used in 3D software to describe the attributes of a 3D object. If you use another software, you can follow the order of this list to enter your own keywords.

The Lightwave attributes are as follows:

Scene < 3D object + Camera + Light

    < Mesh

        < Surface. For example, the Texture Editor has the following parameters: Color, Diffusion, Refraction, Transparency, Emissivity, Alpha-channel, Specularity

        < Texture coordinates

In 3Dmax:

Scene < 3D object + Camera + Light

    < Mesh

      < Material

        < Shader. For example, a Blinn shader has the following parameters: Ambient, Diffuse, Specular, Glossiness, Soften, Opacity, Self-Illumination

        < Textures

In Maya:

Scene < 3D object + Camera + Light

    < Mesh

      < Shader

        < Texture

In Virtools:

Scene < 3D object + Camera + Light

    < Mesh

      < Material

        < Shader. For example, a Gouraud shader has the following parameters: Color, Diffusion, Refraction, Transparency, Emissivity, Alpha-channel, Specularity

        < Texture

In Director 3D:

Scene < Models + Camera + Light

   < Model Resource

      < Shader. For example, a Standard shader has the following parameters: Ambient, Diffuse, Specular, Emissive, Shininess, Blend, Transparent

      < Texture

If you use another 3D software of your choice, enter the information here:

Scene < Models + Camera + Light

   < 3D geometry _____

   < Shader. For example _____

   < Texture _____

The previous tree structures show how each parameter is dependent on another parameter located higher up in the chain. For example, the action of applying a bitmap texture on a 3D object can be described in five steps:

1 The 3D object is made of an arrangement of several meshes.

2 The look and feel of each mesh depends by default on the order in which vertices have been created. This can be altered by creating surfaces or materials on the surface of the mesh.

3 Materials assigned to a 3D mesh can retain information about the choice of a set of texture coordinates or texture maps with planar, cubic, spherical, cylindrical, and UV maps.

4 The texture is applied to the 3D object according to the texture map information provided by the materials attribute.

5 The 3D object is saved with a set of texture coordinates and a path to the file of the bitmap, in this case a 2D image used as a texture.

Rendering a camera view in OpenGL starts from building 3D objects, a camera viewpoint, textures, lights, and environmental effects; it ends with a fully rendered window. The OpenGL pipeline first processes the list of vertices describing the 3D object. Down the pipeline, polygons are built like elements of a pop-up book. Polygons are rasterized by filling the

A character walking on the sand can create a dust ripple, which follows the character's footsteps. The ripple starts around the character's feet and affects pixels further away from the character.

surfaces with fragments of textures. The viewer watches the rendering window where 3D objects appear with interpolation, texturing, and coloring.

Simple demos can help you explore parts of the rendering pipeline. For example, you can compare the lighting effect of a spotlight applied to a single polygon and the lighting effect of the same spotlight applied to a surface divided four times and then divided sixteen times. You will notice that the rendering of the spotlight is more realistic when you increase the number of polygons.

For example, a character walking on the sand can create a dust ripple, which follows the character's footsteps. The ripple starts around the character's feet and affects pixels further away from the character.

This chain of attributes, commonly used with computer animation programs like Pixar's Renderman, is similar to the architecture of the OpenGl pipeline. The animator designing new shaders inside Renderman cannot see an immediate change in the scene without rendering the frame, which can take several minutes, leaving them with a final rendering made of a sequence of rendered frames. This chapter covers examples of interactive shaders, effects applied to a 3D model. The animators can interact in real time with each vertex of a 3D model, which can be programmed with geometric deformation, lighting, or fog attenuation—for example, growing a beard on a character's head at a frame rate of more than $1/30$ of a second. Observing the chain of attributes, we will look at adding scripts to the material attributes to control the texture's behavior.

Each vertex inside a textured 3D object has a five-dimensional set of co-ordinates that are made up of 2D and 3D coordinates. 2D coordinates describe the position of an image in space, and 3D coordinates describe the position of a vertex in space. In the case of the vertex of a 3D object, the vertex receives a 2D set of coordinates describing pixels from a 2D image. Behaviors for interactive textures can change the way pixels are applied to polygons. In this chapter we will cover ways to switch textures, to animate textures and video, or to move them on 3D surfaces.

The rendering pipeline is elegantly designed so you can process vertices with interactive shaders in loops connected to any point of the main stream of the pipeline. Loops for shaders, using simple mathematic operations, are connected to the main traffic of the pipeline without disrupting the core of the rendering process. Coding shaders can be done in Virtools by pro-gramming inside the Shader Editor window and in Director by writing code in the Message window. A set of vertices coordinates is processed inside the loop and returned to the flow of the rendering pipeline as a modified texture image. The amount of programming taking place inside program-mable fragments can affect lighting, environmental mapping, particle systems, fog, nonphotorealistic rendering (toon shading), multiple reflec-tions, refractions (mirrors), bump maps, and vertex animation (mesh transformations).

This chapter covers in depth how to let viewers interact with various kinds of textures. Interactive examples for the tutorial can be found on the com-panion CD-ROM. The examples of texture-based behaviors illustrated in this chapter will allow you to create powerful content without requiring programming skills. The following examples are easily implemented in virtual spaces and can be controlled by the viewer. It is hoped that these examples will challenge you to explore new interactive shaders that may change the way we look at interactive 3D in the near future.

1.3   Displacement Maps

Textures can be used to model 3D objects. The third group of tutorials includes textures used as **displacement maps** to create mesh modifica-tions—for example, textures that can modify a 3D mesh based on the levels of color present in a 2D image. Carefully painted 2D textures can turn a flat mesh into an instant terrain, which would be tedious to model with

Displacement maps combined with vertex rendering can be used as a data visualization tool to monitor the evolution of complex phenomena.

3D software. Displacement maps combined with vertex rendering can be used as a data visualization tool to monitor the evolution of natural phenomena.

We will see an application of displacement maps to create a futuristic city inspired by Ridley Scott's *Blade Runner* in the step-by-step tutorial. This tutorial about real-time displacement maps can be found in the second part of this chapter. More in-depth tutorials about how to use terrains can be found on the companion CD-ROM under Kinematics > Tutorials > cloning.pdf.

1.4   Procedural Painting

The fourth group of tutorials covers several **procedural painting** techniques that can be used inside a virtual space. This group includes the following:

- Vertex rendering, which can place textures in space according to the location of vertices of a 3D mesh

- Painting in 3D, a process that simulates brush strokes covering 3D objects

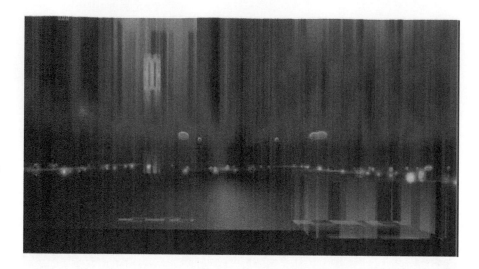

View from the virtual space installation Infinite City by Jean-Marc Gauthier, Miro Kirov, and James Tunick. The virtual space is made of one particle emitter and two textures, which are cloned many times to recreate the complexity of a cityscape.

1.5   Particle Animation

The last group of tutorials is based on particle animation. Particle emitters are designed to push textures living for a determined time inside a virtual space. The following example shows an urban virtual space generated by a single-point particle emitter and two textures scrolling on four-sided polygons. Particle emitters use still images, videos, and movies to texture map hundreds or thousands of particles made of 3D sprites. Particle animation systems can create clouds of 3D billboards flying in the same direction across a virtual space. Each billboard is made of a single polygon textured with a still image or with a sequence of frames from a movie.

2   TEXTURE PRIMER: CREATING A SKYBOX

2.1   Origins of the Skybox

A skybox is a cube with 360-degree views of landscape that can be used as a 3D background for a virtual scene. Skyboxes follow the tradition of panoramas, which were built before movies were invented. Panoramas offered views of landscapes and battle scenes to a paying audience. Early examples of panoramas used a cubic volume similar to the inside space needed for a small theater. The audience of 10 to 12 people would sit at one end of a decorated stage shaped like a shoe box. Changes were made inside the "shoe box" in front of the naked eye of the audience by changing lighting and moving backgrounds. Scrolling or rotating paintings on canvas were commonly used to suggest a transition from day to night, a change of season, or clouds moving in the sky. Panorama entrepreneurs such as Daguerre, the inventor of photography, were carefully covering all walls with seamless images so as to maintain the illusion of the scene and not reveal the angles or limitations of a room. Later examples of panoramas show that cylinders and spherical domes were used as skyboxes.

Similar seamless textures used as backgrounds for virtual worlds can be applied the same way inside cubes, spheres, or cylinders. The inside face of a cube can be textured with images that tile perfectly with other faces sharing a common side. The skybox textures cover all the inside faces of a cube the same way wallpaper can be applied to the walls, floor, and ceiling inside a room. The seamless skybox textures inserted inside the cube provide a continuity of background regardless of the camera's motions.

Skyboxes are large-size cubes that may contain a virtual world 10 times smaller. Realistic bitmaps for skybox textures integrate lighting elements and environmental effects from a virtual world. For example, sunlight, moon, and clouds from a scene can be integrated into a landscape viewed at sunset displayed on the skybox. We will show how to create this in Skytracer, Lightwave's procedural textures generator for skies. Please note that Virtools provides a Sky Around building block, a procedural skybox behavior that is not used for this tutorial.

Realistic bitmaps for skybox textures integrating sunlight, moon, and clouds from a scene can be generated in Lightwave Skytracer, Lightwave's procedural sky textures generator.

2.2   Skybox Tutorial

This tutorial is a basic introduction to modeling a cube with textures. If you are using Maya, please refer to Interactive Textures > Tutorials > cube-Maya.pdf on the companion CD-ROM to make a polygonal skybox based on a cube. For instructions on making a spherical skybox based on NURBs, please refer to Textures > Tutorials > sphere-Maya.pdf. We will reuse the textured 3D model of the cube in other sections of this chapter. The cube will be covered with interactive textures. Although the tutorial covers step by step how to use textures in Lightwave, you can decide to go straight to Section 3 if you already have an understanding of creating textures in a noninteractive environment.

The first part of this tutorial covers the simple steps taken to model and texture a cube that we will use as a skybox. The skybox, created in Light-wave, is exported in Virtools, where it is used for examples of interactive textures covered in the next section of this tutorial. You can find the files for this tutorial on the companion CD-ROM under Interactive Textures. You can also bypass this tutorial and go directly to the textures examples described in the next section.

Creating a skybox with interactive textures will allow you to position default textures that later can be replaced by other textures or videos of the same size.

Although setting up texture coordinates is necessary before adding inter-activity, textures can also be created inside Virtools.

2.3   Design of a Skybox in Lightwave Modeler

Step 1. Open Lightwave Modeler, and select Create > Object > Box. Drag the blue markers to draw a cube.

Step 2. Select the blue markers in another view. Drag to draw the cube. Hit the space bar to confirm the creation of the cube.

Step 3. Select Polygons. Click on one of the cube's polygons. Go to Detail > Surface. In the Change Surface window, create surface name and color. Repeat the previous step to assign a surface name and color to each side of the cube.

Step 4. In Surface Editor, select surface name, and go to Basic > Color > T.

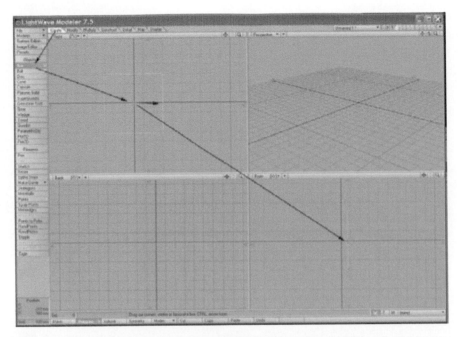

Step 1. Open Lightwave Modeler. Select Create > Object > Box. Drag the blue markers to draw a cube.

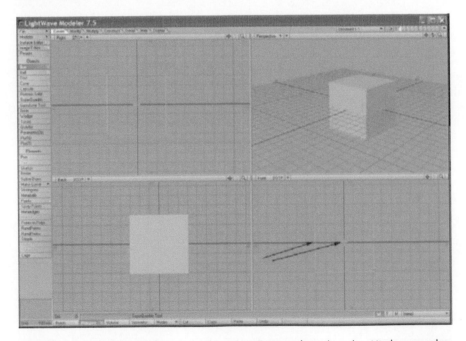

Step 2. Select the blue markers in another view. Drag to draw the cube. Hit the space bar to confirm the creation of the cube.

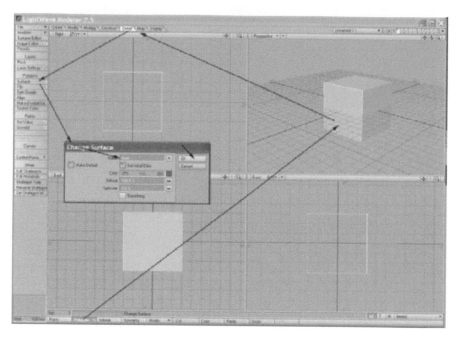

Step 3. Select Polygons. Click on one of the cube's polygons. Go to Detail > Surface. In the Change Surface window, create surface name and color.

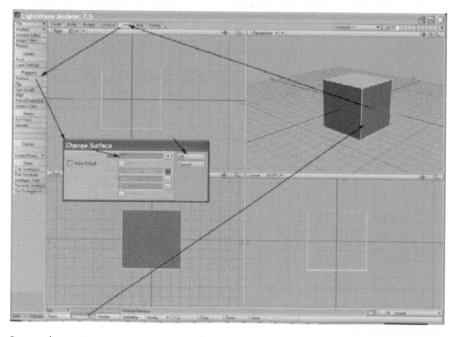

Repeat the previous step to assign a surface name and color to each side of the cube.

The Texture Editor opens. Then select the surface name > Axis > Automatic Sizing > Use Texture. Please note that all textures should be saved as .jpg format images in Photoshop® or in the image-processing software of your choice. Repeat the same operation for each surface until you cover the whole cube with a group of images.

Step 5. Select Polygons in the lower left of the screen. Select all the cube's polygons. Right click the mouse to draw a lasso around a wireframe preview of the cube. Go to Details > Polygons. Flip normals (yellow lines) sticking out of all the selected polygons of the cube. Normals of the cube faces and textured images are now pointing toward the inside of the cube.

Step 6. Select the first layer halfway. In the second layer, create a second cube larger than the first one on

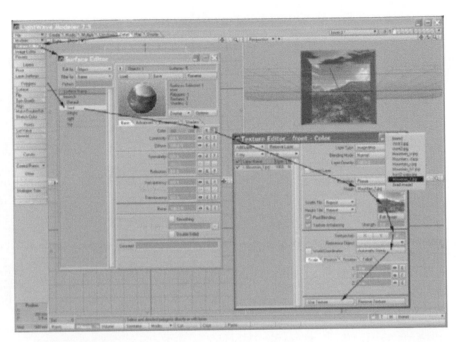

In Surface Editor, select surface name. Go to Basic > Color > T.

Repeat the same operation for each surface until you cover the whole cube with a group of images.

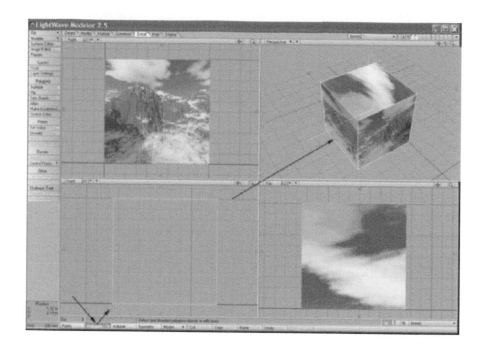

Select Polygons in the lower left of the screen. Select all the cube's polygons. Right click the mouse to draw a lasso around a wireframe preview of the cube.

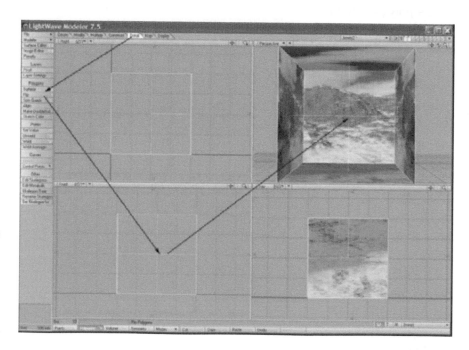

Go to Details > Polygons. Flip normals (yellow lines) sticking out of all the selected polygons of the cube. Normals of the cube faces and textured images are now pointing toward the inside of the cube.

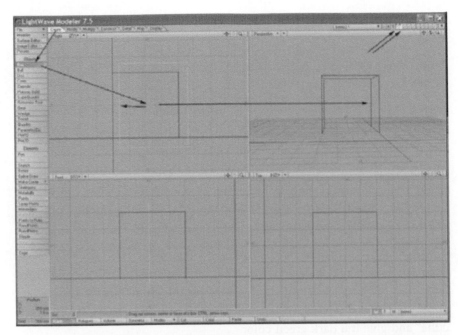

Select the first layer halfway. In the second layer, create a second cube larger than the first one on all the sides except the front. We will use the opening to pick inside the cube.

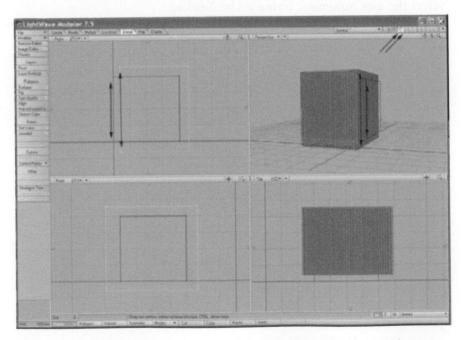

Adjust the size of the second cube in green relative to the first one in black wireframe.

all the sides except the front. We will use the opening to pick inside the cube.

Step 7. Adjust the size of the second cube in green relative to the first one in black wireframe.

Step 8. Go to Multiple > Boolean > Subtract. Select the first layer, go to a wireframe mode viewport, and right click and drag the mouse to select all the polygons. Delete the small cube from the scene.

Step 9. Add a new texture to the second green cube. Take a screen capture of the front view by clicking Print Screen on your keyboard. In Photoshop®, go to File > New > Edit > Paste. Crop the picture along the outside edge of the second green cube. Paint the texture, and save the image as "text1.jpg".

Step 10. In Modeler assign a surface to the front side of the second green cube. Go to Polygons (lower left). Click on the front face of the second green cube. Go to Detail > Map > Surface. Name the surface "front-green cube" and give it a dark green color.

Step 11. Go to Surface Editor, and select the surface name "front-green cube." Go to Basic > Color > T. In the Surface Editor, select "text1.jpg." Select the axis Automatic Sizing > Use Texture.

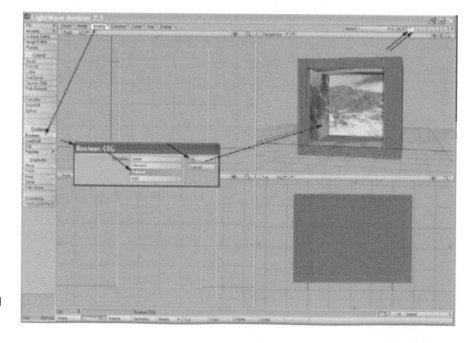

Go to Multiple > Boolean > Subtract. Select the first layer. Go to a wireframe mode viewport, and right click and drag the mouse to select all the polygons. Delete the small cube from the scene.

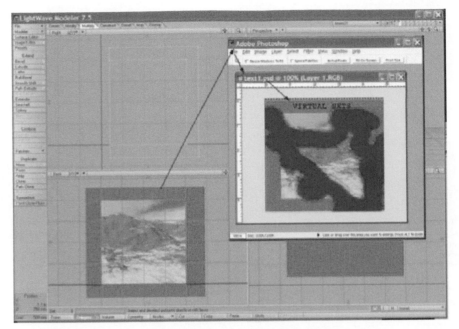

Add a new texture to the second green cube. Take a screen capture of the front view by clicking Print Screen on your keyboard. In Photoshop®, go to File > New > Edit > Paste. Crop the picture along the outside edge of the second green cube. Paint the texture and save the image as "text1.Jpg."

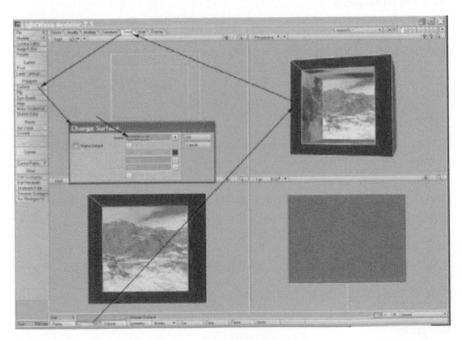

In Modeler assign a surface to the front side of the second green cube. Go to Polygons (lower left). Click on the front face of the second green cube. Go to Detail > Map > Surface. Name the surface "front-green cube" and give it a dark green color.

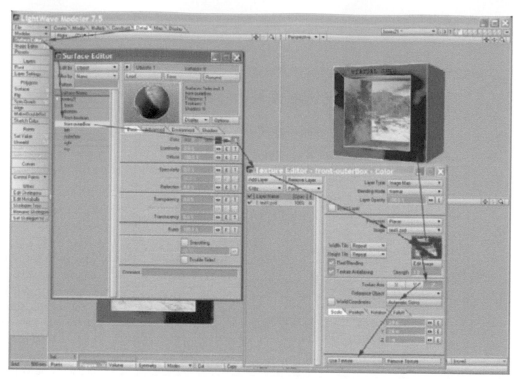

Go to Surface Editor, and select the surface name "front-green cube." Go to Basic > Color > T. In the Surface Editor, select "text1.jpg." Select the axis Automatic Sizing > Use Texture.

Step 12. Go to Construct > Triple. The model is now composed of triangles or three point polygons. Save your file. Go to File > Save Object. Save as "skybox.lwo" (lwo stands for Lightwave object).

2.4　Create a Scene with Lights and Cameras in Lightwave

Step 1. Go to File > Load Object. Open "skybox.lwo" from the Interactive Textures folder on the companion CD-ROM. Please note that you may find that the center of the object needs to be recentered. Go to Items > ToolsMore > Pivot > Move Pivot Point Tool. Drag the arrows until the pivot point is aligned with the center of the cube.

Step 2. Go to Items > Motion Options > Target Item > Select the Cube. The camera is now locked to a new target—the pivot point of the object.

Step 3. Select the light. Go to Items Properties. > In Light Properties change the light's color. Repeat the previous step to target the spotlight on the cube.

Step 4. Go to File > Save Scene As, and save as "skybox_scene.lws" (a Lightwave scene).

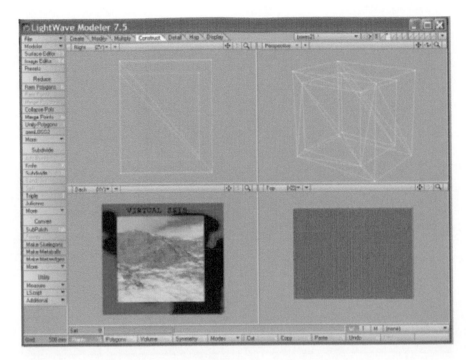

Go to Construct > Triple. The model is now composed of triangles or three-point polygons.

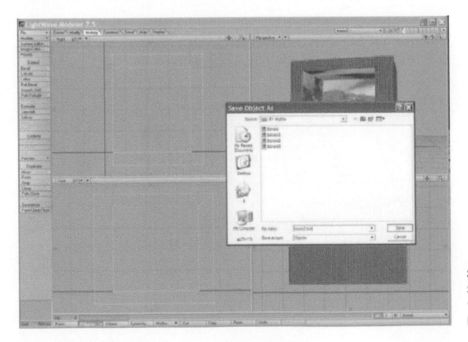

Save your file. Go to File > Save Object. Save as "skybox.lwo" (lwo stands for Lightwave object).

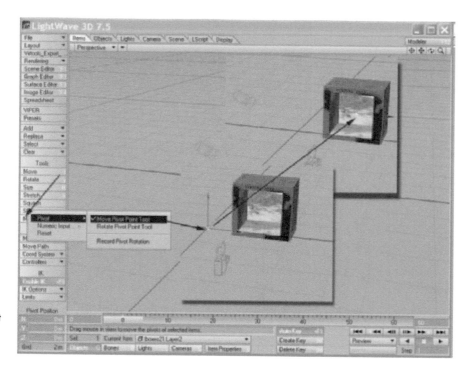

Go to Items > Tools—More >
Pivot > Move Pivot Point Tool.
Drag the arrows until the pivot
point is aligned with the
center of the cube.

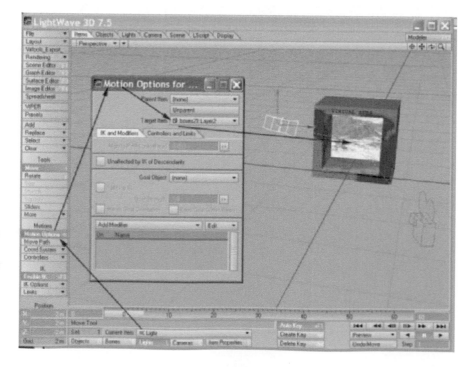

Select the camera by clicking
on it inside the viewport or in
the Current Item list at the
bottom of the interface. Go to
Items > Motion Options >
Target Item. Select the cube.
The camera is now locked to
a new target—the pivot point
of the object.

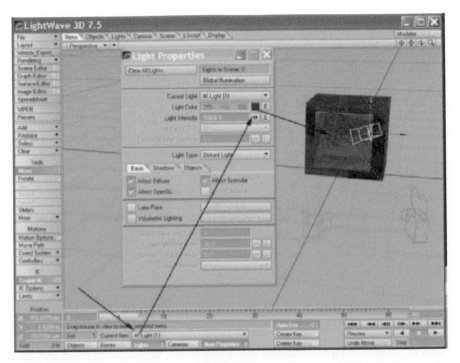

Select the light. Go to Items Properties. In Light Properties change the light's color. Repeat the previous step to target the spotlight on the cube.

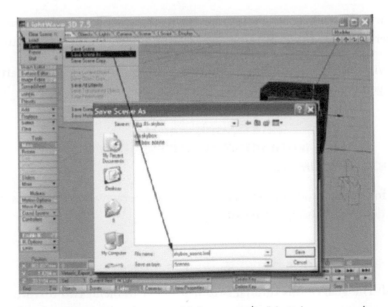

Go to File > Save Scene as, and save as "skybox_scene.lws" (a Lightwave scene).

2.5   Creating a Database to Manage 3D Assets in Virtools

Step 1. In Virtools, go to Create New Data Resources. This will create a skybox.rsc file, which can open an asset folder.

Step 2. Please make sure that all texture images in .jpg format are available in the Skybox > Textures format. Copy and paste the texture files in the Textures folder as needed.

Step 3. Export the scene from Lightwave to Virtools. Go to Items > Virtools Export, and select OK. Save the object in the 3D Entities folder.

Step 4. In Virtools go to Resources > Open Data Resource, and select skybox.rsc. Open skybox.rsc from the Interactive Textures folder on the companion CD-ROM. The asset folders are visible on the top right of the screen. Select 3D Entities; the skybox file is visible.

Step 5. Go to Skybox Data on the upper right. Select 3D Entities. Click and drag the Skybox_model file into the Layout window.

Step 6. Toggle between Camera and Perspective view. The camera and the lighting are now visible. Arrows located in the lower part of the screen show the relationships between Layout and the listing of 3D assets under Level Manager.

In Virtools 3D Layout, choose the point of view of the camera created earlier in Lightwave in the upper middle pull-down menu for Layout.

The view in 3D Layout appears with textures, lighting, and camera settings that you created in Lightwave.

3   CREATE INTERACTIVE TEXTURES: BITMAPS AND PROCEDURAL IMAGES

The following tutorials cover various kinds of interactive textures presented in order of increasing complexity.

3.1   Blending Textures Looped in Time

Goal: The goal of this tutorial is to learn how to blend two textures over time.

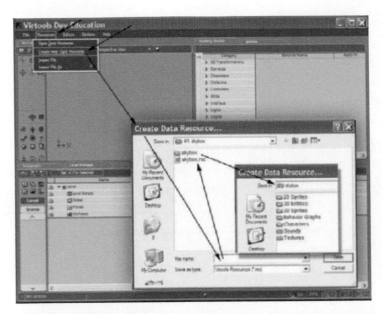

In Virtools, go to Create New Data Resources. This will create a skybox.rsc file, which can open an asset folder.

Please make sure that all texture images in .jpg format are available in the Skybox > Textures format. Copy and paste the texture files in the Textures folder as needed.

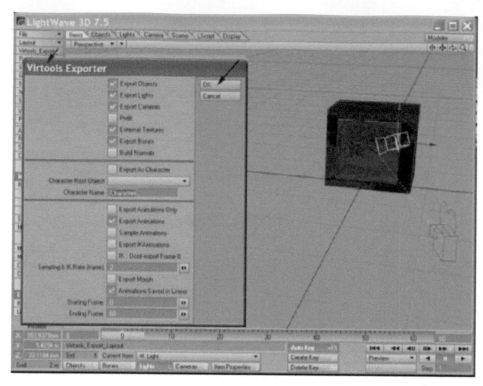

Export the scene in Lightwave. Go to Items > Virtools Export, and select OK.

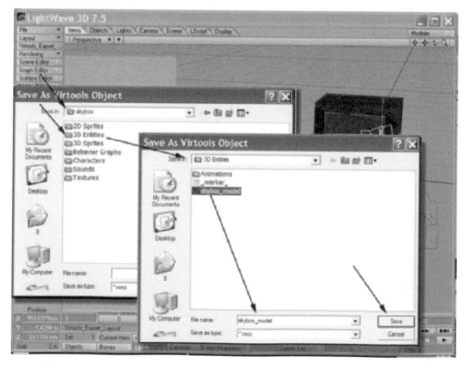

Save the object in the 3D Entities folder.

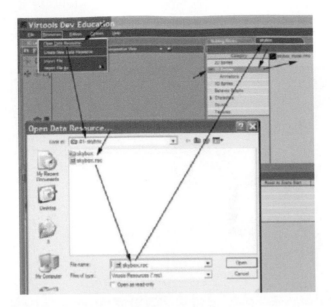

In Virtools go to Resources > Open Data Resource. Select skybox.rsc. The asset folders are visible on the top right of the screen. Select 3D Entities; the skybox file is visible.

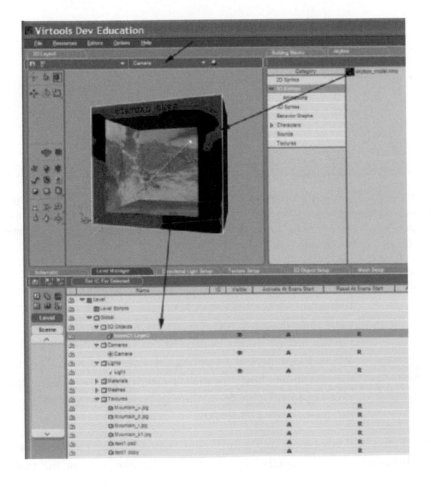

Go to Skybox Data on the upper right, and select 3D Entities. Click and drag the Skybox_model file into the layout window. Choose the point of view of the camera created in Lightwave in the upper-middle pull-down menu for Layout. The view appears with the same textures, lighting, and camera settings created in Lightwave.

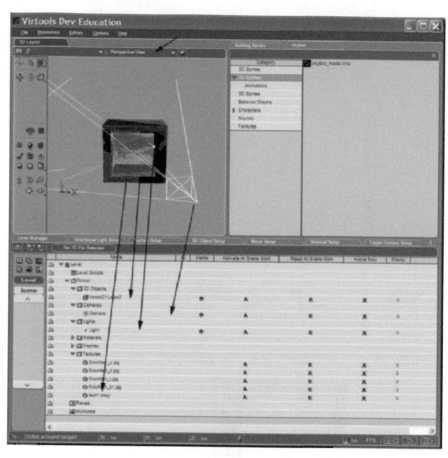

Open skybox.cmo from the Interactive Textures folder on the companion CD-ROM. Toggle between Camera and Perspective view. The camera and the lighting are now visible. Arrows located in the lower part of the screen show the relationships between Layout and the listing of 3D assets under Level Manager.

A timer can control intensity and duration of the Texture Blending effect.

Overview: This tutorial shows how to control the fade-in and fade-out between two textures over time. One texture changes gradually into the other and reverses back into itself. A timer can control intensity and duration of the Texture Blending effect. We will look here at repeating the effect in a loop. The next tutorial shows how the viewer can trigger the effect. Open textures-blend.cmo from the Interactive Textures folder on the companion CD-ROM.

The Times building block, step 1 in the illustrations, triggers an iteration of the behavior every 6 seconds. The building block takes care of the transi-

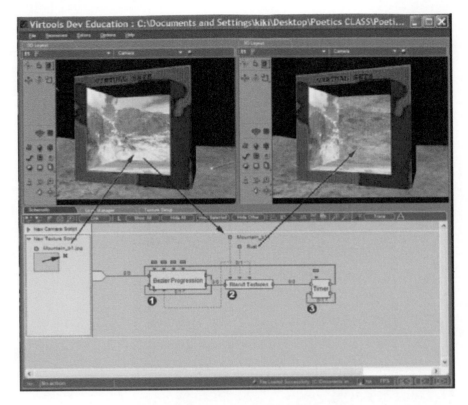

The Bezier curve or Bezier Progression building block, noted in step 2 in the illustrations, is an interactive way to control the intensity of the blending.

tion over time between the textures. Two time controllers are used in the looped behavior. The Blending Textures building block, step 2 in the illustrations, is a script for a procedural texture that can process up to two more textures in addition to the existing texture.

The Bezier Curve or Bezier Progression building block, noted in step 2 in the illustrations, is an interactive way to control the intensity of the blending. The bell-shaped curve lasts 5 seconds on the X-axis timeline and varies in intensity on the Y axis. The curve starts at time zero and at zero value for blending. It reaches a maximum value of blending at 2.5 seconds and returns to zero value at 5 seconds. You can edit the curve by inserting and dragging control points.

How to use it? Texture blending is a discreet but powerful way to create subtle changes of mood in a virtual scene. In the Futuristic City example in section 9 we will blend textures, from color to black-and-white, to simulate the transition from day to night. A gradual change from daytime

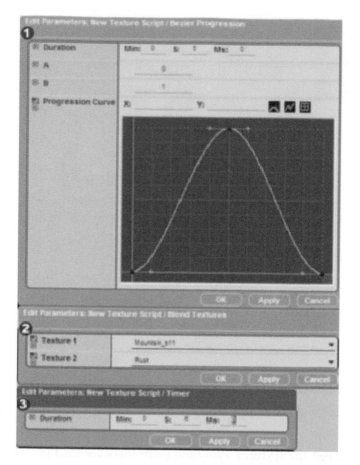

The bell-shaped curve lasts 5 seconds on the X-axis timeline and varies in intensity on the Y axis. The curve starts at time zero and at zero value for no blending. It reaches a maximum value of blending at 2.5 seconds and returns to zero value at 5 seconds.

A gradual change from daytime textures with natural lighting to a nighttime scene with artificial light can contribute to the viewer's suspension of disbelief.

textures with natural lighting to a nighttime scene with artificial light can contribute to the viewer's suspension of disbelief.

3.2   Blending Textures Controlled by Viewer's Input

Goal: The goal of this tutorial is to learn how to script two textures so they can blend—fade-in and fade-out—when pressing a keyboard key.

Overview: This tutorial shows how a viewer can trigger the effect to control the fade-in and fade-out between two textures. A viewer can control intensity and duration of the Texture Blending effect. We will look at an example of a blending effect with set parameters, texture-blending-viewer.cmo.

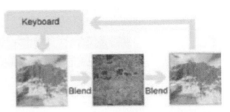

A viewer can control intensity and duration of the Texture Blending effect.

This tutorial shows several ways to control the transformation of a texture using the keyboard. When pressing the 1 key on the keyboard, one texture seems to get brighter. In this case the texture changes gradually into the brighter copy of it and reverses back into itself. When pressing the 2 key, the same texture changes gradually into a sky image and reverses back into itself. The Switch on Key building block, step 1 in the illustrations, allows you to rig the keyboard keys to the procedural texture and to control the changes from one texture to the other. Blending Texture and Bezier Progression building blocks have been encapsulated in one new

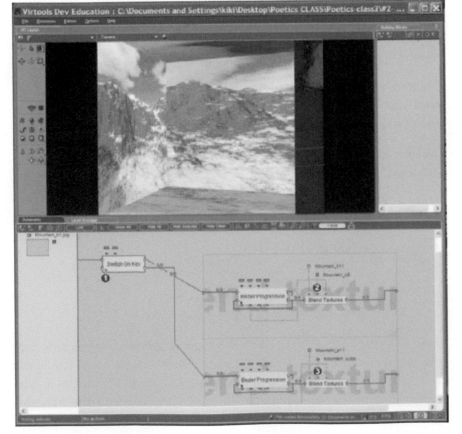

The Switch on Key building block, step 1, allows you to rig the keyboard keys to the procedural texture and to control the changes from one texture to the other. Several duplicates of the Blend Texture behavior, steps 2 and 3 in the illustrations, are linked to out pins of the Switch on Key building block called "out 1" and "out 2".

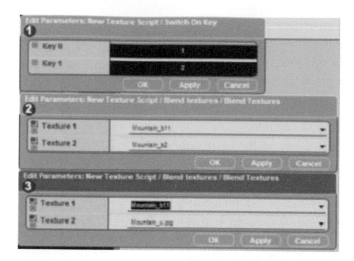

Parameters for Switch on Key and Blend Textures building blocks.

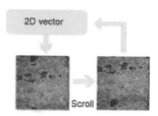

The texture is scrolling vertically.

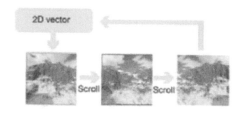

The texture is scrolling horizontally.

behavior called Blend Texture. Several duplicates of the Blend Texture behavior, steps 2 and 3 in the illustrations, are linked to out pins of the Switch on Key building block called "out 1" and "out 2," step 1 in the illustrations.

3.3   Scrolling Textures

Goal: This tutorial shows how to move textures across the surface of an object. You can assign several scrolling directions for each texture, such as vertical or horizontal scrolling. This behavior can create very interesting cinematic effects. For example, you can scroll text on a transparent background or scroll a long panoramic image that may be much larger than the object itself.

The texture is scrolling vertically. In this example the surface of the 3D object is the same size as the image used for the texture.

A texture scrolling horizontally is similar to the effect of panning a camera inside a panoramic image. In this example the surface of the 3D object is much smaller than the image used as a texture.

Overview: The Scrolling Texture building block is applied to the object. Textures follow an X,Y direction or a 2D scrolling vector, which are the surface coordinates of the object. The choice of X,Y orientation for the vector determines the direction of the scrolling movement. The length of the vector controls the speed of the scrolling.

3.4   Scrolling Texture with a Transparent Background

Goal: The goal of this tutorial is to show that a texture can be made partially transparent by color-keying-out surfaces of the same color. This effect can be applied to the background of a texture. Color-keyed images or text seem to float in the air without reference to the edges of the polygons holding the texture.

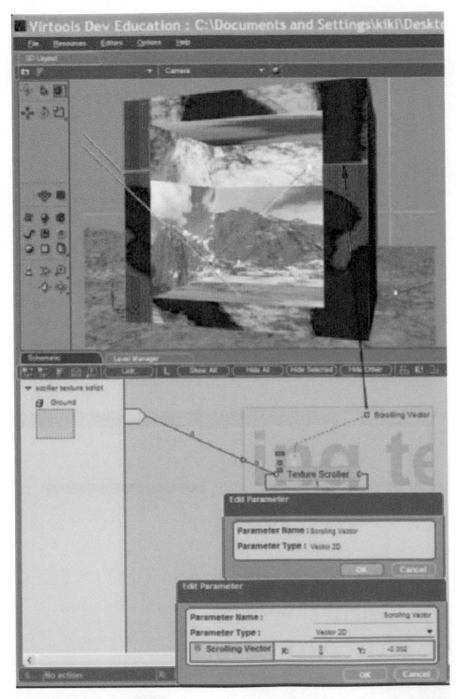

The choice of X,Y orientation for the vector determines the direction of the scrolling movement. The length of the vector controls the speed of the scrolling.

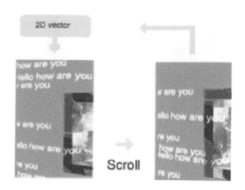

Scroll

Vertical scrolling of text with a transparent background.

Overview: Vertical scrolling of text with a transparent background can be achieved in two steps. The text is laid out on a uniform background color and the picture of the text is color-keyed-out to remove the background. In this example, the text and background were created in Photoshop®. Open texture-scrolling-text.cmo from the Interactive Textures folder on the companion CD-ROM.

The texture, saved in Virtools, Textures Data Resource Folder > Textures, is imported—dragged and dropped into the Virtools 3D Layout window. The imported texture shows under Level Manager > Global > Textures > Text.

Double-click on the texture to open the Texture Setup window. Select Pick Color and Color Key Transparency. The illustration shows two versions of the same text. Step 1 is a color-keyed green text with the background removed. The text seems to float in the air and scrolls vertically like movie credits. Step 2 shows the same text with its original black background.

How to use it? Only two scrolling textures on semitransparent backgrounds were used in the installation Infinite City. The top image shows the origi-

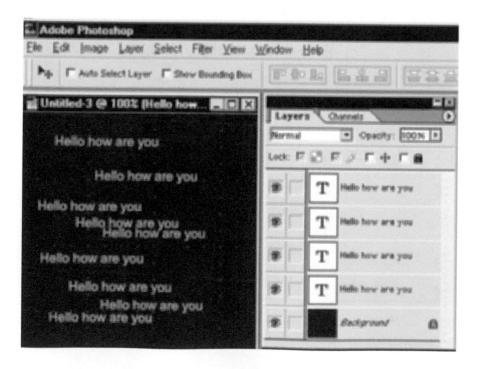

The text and background are created in Photoshop®.

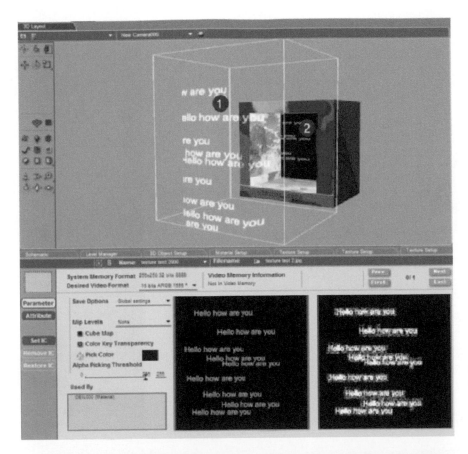

Step 1 is a color-keyed green text with the background removed. Step 2 shows the same text with its original black background.

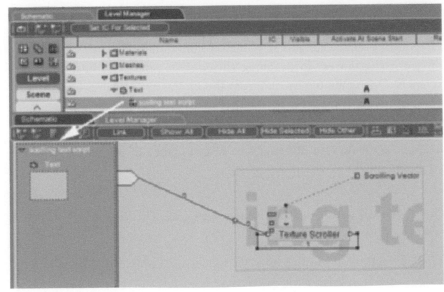

Level Manager and schematic views showing the scrolling text script attached to the texture called Text.

Only two scrolling textures on semitransparent backgrounds were used to create the mood of the installation Infinite City created by Jean-Marc Gauthier, Miro Kirov, and James Tunick.

nal texture; the bottom image shows the alpha-channel black-and-white gradient used to create transparencies. The texture on the left creates the effect of flashing headlights scrolling on horizontal billboards. The texture on the right moves slowly along the vertical shaft of transparent skyscrapers. Open infinite-city.cmo from the Interactive Textures folder on the companion CD-ROM.

Until now, we have been using single images as a texture. In the next example we will set a 3D model with a sequence of images organized as an animation.

3.5 Playing Movie Clips in a Loop Controlled by the Viewer

Goal: This tutorial shows how a 3D model can be used to display movies, video clips, or animations as textures.

Overview: Because a movie is made of a sequence of frames following each other, the correct placement of the first image of the video clip on a 3D model will determine the texture mapping coordinates for the whole clip. Open texture-movie.cmo from the Interactive Textures folder on the companion CD-ROM.

In the 3D software of your choice, the process of texture mapping a movie on a 3D object starts with the same steps as texture mapping a single image.

Create a surface with a texture map made of the first frame of the movie. The correct placement of this texture on the surface of a 3D model will guide the placement in the other frames of the sequence.

In Virtools, you can add a script to the material to swap the initial texture with a new texture, a movie. The script includes two building blocks. The Set Texture building block assigns a new texture to the material with the movie's name. The Movie Player building block controls the playback parameters for the movie.

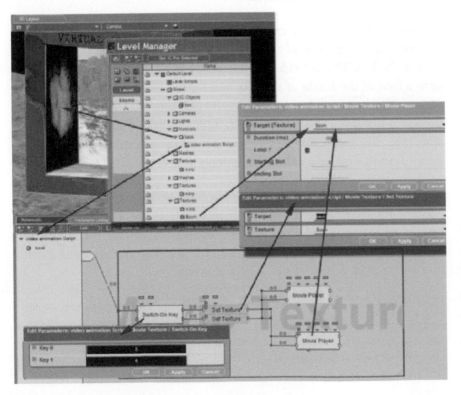

The Set Texture building block assigns a new texture to the material with the movie's name. The Movie Player building block controls the playback parameters for the movie. This setup shows how to swap between movies by using a Switch on Key building block.

3.6   Playing Movie Clips Controlled by the Viewer

Goal: This tutorial shows how the viewer can control the display of movies, video clips, or animations as textures.

Overview: Several movie textures similar to the one described in the previous section are controlled from the keyboard. Open texture-movie.cmo from the Interactive Textures folder on the companion CD-ROM. This setup shows how to swap between movie textures by using a Switch on Key building block. Please note you need to stop the first movie player in order to start playing the second one. This constraint is visible in the arrangement of links between Set Texture and Movie Player building blocks.

Render Top Camera

Creating a texture with an image filmed by a virtual camera.

3.7   Using a Live Video from a Virtual Camera

Goal: This tutorial shows how to create a texture with an image filmed by a virtual camera.

Overview: In this typical case study of a picture in a picture, a virtual camera films a top view of the skybox and a second camera films a front view of the box. The live video feed from the top camera is texture mapped onto a face of the inside of the box and on the ground. Any new element entering the scene will be captured by the top camera and will appear inside the video texture. The second camera films the scene and the 3D objects with their live video textures. Open texture-camera.cmo from the Interactive Textures folder on the companion CD-ROM.

The Texture Render building block creates an additional view, a rendering window, for the images filmed by the top camera. This live video feed can be applied to a group of surfaces inside the scene. The live video filmed by the top camera can also be found inside the small rendering frame located on the top-right corner of the 3D Layout window.

How to use it? A top view of a character walking on a terrain can be very helpful for the viewer navigating the character. An additional navigation window with a view from the top camera can be integrated inside the viewer's interface. In other cases, a "picture in a picture" effect can turn a still life into a virtual world of poetic associations.

The Texture Render building block creates an additional view, a rendering window, for the images filmed by the top camera. This live video feed can be applied to a group of surfaces inside the scene.

A "picture in a picture" effect can turn a simple object into a virtual world of poetic associations.

3.8   Texturing with a Random Live Video Signal

Goal: The goal of this tutorial is to show how to use a self-generated video signal as a texture.

Overview: The texture is processed from a live video signal created from a random noise. The processing of the video signal is always different. The advantage of using random live video signals is being able to display rich textures without the memory overhead inherent to retrieving video footage stored on the hard drive. Open livevideo1.cmo from the Interactive Textures folder on the companion CD-ROM.

The script for the interactive video texture uses a Write in Texture building block to write with a Pen Texture or a colored pixel inside a texture. In this example, the Pen Texture is using the Background Script. Two Random building blocks are used to generate unpredictable X,Y positions for the Pen Texture.

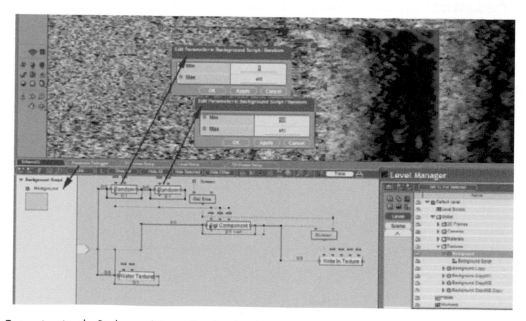

The Pen Texture is using the Background Script. Two Random building blocks are used to generate unpredictable X,Y positions for the Pen Texture.

Live video signals replacing still images can add a theatrical dimension to a virtual set.

How to use it? Live video signals replacing still images can add a theatrical dimension to a virtual set. In this example, the live video signals replace original textures of 3D objects.

3.9　Texturing with Live Video Signals Controlled by the Mouse

Goal: The goal of this tutorial is to show how to process textures from a live video signal that move according to the viewer's mouse.

Overview: The Write in Texture building block, step 2, uses a texture or a colored pixel inside a texture. Note the ripples created by the damping effect. In this example, the Pen Texture is using the Background Script. Step 1, the Get Mouse Position building block, controls the movement of the texture across the surface. Step 2 creates a moving image. Step 3 processes the damping effect on the edges of the surface. Open livevideo2.cmo from the Interactive Textures folder on the companion CD-ROM.

3.10　Controlling a Texture with an Animated Character

Goal: The goal of this tutorial is to show how to process a live video signal following an interactive character.

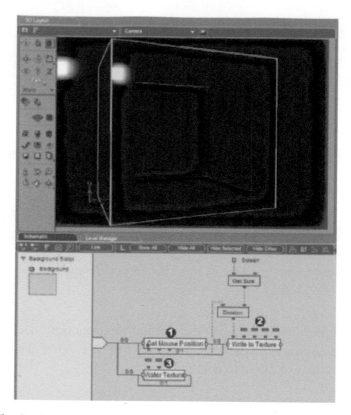

Step 1. The Get Mouse Position building block controls the movement of the texture across the surface. Step 2 creates a moving image. Step 3 processes the damping effect on the edges of the surface.

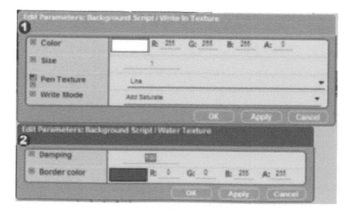

Parameters for the Write in Texture and Water Texture building blocks.

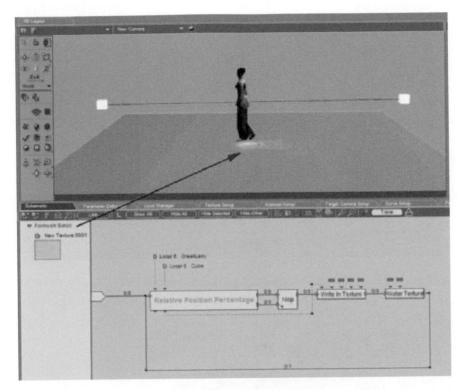

The Relative Position Percentage Script is calculating the position of a character walking along the red path relative to the size of the blue floor. The value is passed to the Write in Texture building block.

Overview: The Relative Position Percentage Script is calculating the position of a character walking along the red path relative to the size of the blue floor. The value is passed to the Write in Texture building block. Please refer to the Overview in Tutorial 3.9 for more details.

How to use it? Dust, ripples, and shadows are some of the effects following a character's footsteps. The texture effect is linked to the path of an animated character. This effect can reinforce the presence of large crowds of cloned characters on the ground. The following illustration shows interferences of ripples following several characters. Open texture-ripple.cmo from the Interactive Textures folder on the companion CD-ROM.

3.11   Texturing with Two Distinct Live Video Signals

Goal: The goal of this tutorial is learning to process two textures from two live video signals.

The ripple effect on the ground can reinforce the presence of multiple characters.

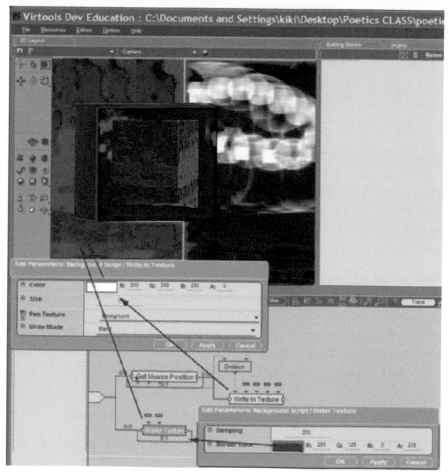

The first texture has a blue border color, and the second has an orange border color. The second texture is duplicated from the original live video texture.

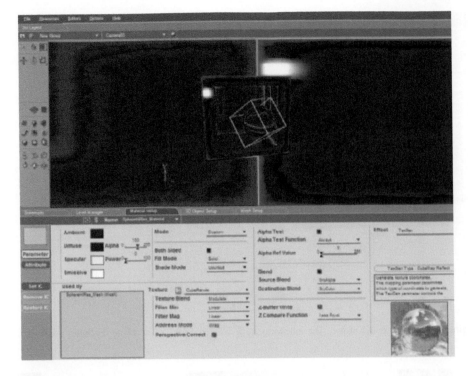

A spherical mirror is placed at the center of the frame.

Overview: The first texture has a blue border color and the second has an orange border color. The second texture is duplicated from the original live video texture. Each texture is applied to a distinct surface inside the box. Open 2videos.cmo from the Interactive Textures folder on the companion CD-ROM. Please refer to the Overview in Tutorial 3.9 for more details.

How to use it? The following example is a setup that can be used for an interactive screen. The dynamic composition shows textures overlapping each other in a 3D space controlled by sound. A spherical mirror is placed at the center of the frame.

3.12 Texturing with Two Live Video Signals Reflected in a Mirror

Goal: The goal of this tutorial is to turn the surface of an object into a reflective mirror. Open texture-mirror.cmo from the Interactive Textures folder on the companion CD-ROM.

Overview: This shader is added to the original material of a 3D object. The following steps will take you to the Material Setup window. You can

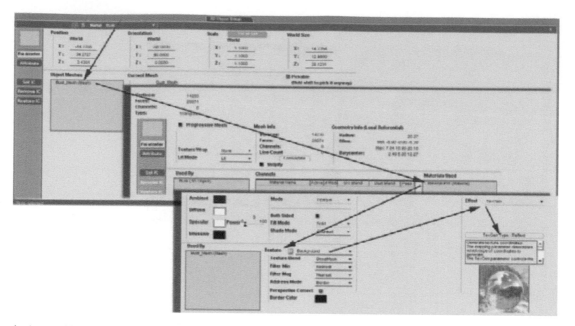

This shader is added to the original material of a 3D object. The following steps will take you to the Material Setup window.

Replacing an original texture with reflections from the environment.

double-click on the object in 3D Layout and select Setup or double-click on the object's name in the Level Manager window. Inside the 3D Object Setup window, double-click on the object's mesh. Inside the Material Setup window, select Effect. Choose TexGen > Reflect.

How to use it? This dynamic effect can change the appearance of a 3D object by replacing an original texture with reflections from the environment.

3.13 Texturing with Live Video Stream from a Video Camera

Goal: The goal of this tutorial is to show how to create textures on a 3D object with a live video stream from a USB video camera.

Overview: A video camera plugged into the USB port of your computer can stream live video images, which are used as textures on the surface of a 3D object.

How to use it? Real-time video images commonly are used in interactive television installations and commercial applications where the viewer can take his or her own picture or pictures of the viewer's physical environment and use it as a texture on a virtual character or on billboards inside the virtual world. Open webcam.vmo from the Interactive Textures folder on the companion CD-ROM. Please note that the SIALogicsWP.dll needs to be installed in the player's Building Blocks folder before testing the Webcam inside the virtual world. The SIALOgics.dll for streaming a Webcam inside a virtual world is available at www.siasistemas.com.

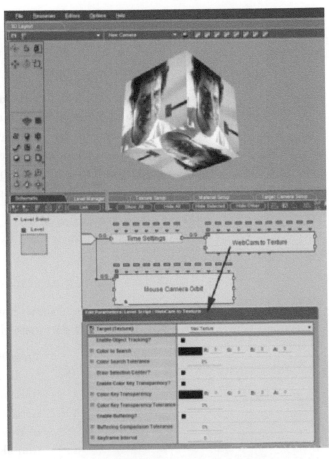

Live video streamed from a USB video camera is textured onto a 3D cube.

4   PROGRAMMABLE SHADERS

The shaders described in this section are available for users of DirectX 9 and users of graphics cards compatible with VS1.0 shader.

The following shaders are being developed as part of the Dynamic Virtual Patient (DVP) project, in collaboration with Martin Nachbar, head of Advanced Educational Systems at New York University's College of Medicine. The project is looking at dynamic simulations of a virtual patient for training of students and surgeons. The DVP is one large interface that can run simulations between body parts and organs connected together. When viewers "touch" a virtual organ with a virtual hand, they turn on a dynamic chain of connections, actions, and reactions connecting other organs. See www.tinkering.net/virtualpatient for more information about this project.

The following shaders are created as part of the DVP project. We are researching ways to display X-rays of areas of a virtual body directly located under the virtual camera. The X-ray shader can alter the rendering of 3D objects and let the viewer see through the complex architecture of the human body. The depth of field shader combines renderings of an object with its own alpha-channel to render an image similar to a CAT scan of a 3D model of the body.

Coded shaders can be associated with the material of a 3D object. The association of the material with a shader can override or replace a texture previously assigned to the material. Virtools provides a Shader Editor where you can write your own code including descriptions of vertex shaders, pixel shaders, and shading techniques. After compiling the code, shaders can be edited in real-time for changes of parameters, shader techniques, or types of shaders.

Let's set up a shader step by step.

4.1   The X-Ray Shader, Created by Zach Rosen

Goal: This tutorial shows how to set up a shader that can deliver an instant X-ray of a body part located inside a virtual body.

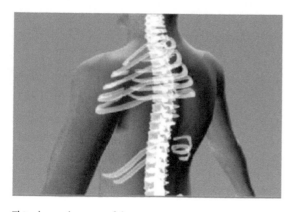

Overview: The following examples present several applications of shaders for medical visualization. The following illustration shows the powerful rendering possibilities of the X-ray shader applied to the bones of the thorax of a virtual patient. The next illustration shows the structure of the upper body visible though the skin. In this case, the X-ray shader is applied to the skin covering the neck and face. Open shader-xray.cmo from the Interactive Textures folder on the companion CD-ROM.

This shows the powerful rendering possibilities of the X-ray shader applied to the bones of the thorax of a virtual patient. 3D model by Miro Kirov, AES. Shader by Zach Rosen.

Let's create a shader in the Shader Editor:

Step 1. Go to the Shader Editor and open a new window. You can edit existing code from Virtools using the default code or you can create your own code from scratch.

Vertex shaders are computed first—for example in the following code:

```
//---Basic vertex transform

    float4 screenPos = mul(float4(Pos, 1), Wvp);

    Out.Pos = screenPos;</EXT>
```

Pixel shader functions (for example, in the following code), reuse the previous variables:

```
//---Pixel shader function

float4 FrontPS(VS_OUTPUT PSIn): COLOR

    float4 result = frontColor;

    float4 texCol = tex2D(texSampler,PSIn.
    TexCoord);

    float texAlpha = dot(texCol,1)/3;

    result[3] = clamp((PSIn.Opacity+ texAl-
    pha)/2.1,0,1);

    return result;
```

This illustration shows the structure of the upper body visible though the skin. In this case, the X-ray shader is applied to the skin covering the neck and face. 3D model by Miro Kirov, AES. Shader by Zach Rosen.

You can select manual parameters, outlined in the following code, as front color, back color, and density.

```
//---Manual Parameters

float density = 0.4;

float4 frontColor = {1, 1, 1, 1};

float4 backColor = {.5, .5, .5, .5};</EX>
```

These parameters will remain accessible after you compile the code.

In any case you will need to reuse some of the variables and parameters present in the default Virtools shader. The editor lets you combine vertex shaders and pixel shaders. Press Compile to create a new shader available in the right column.

Zach Rosen created the following code for the X-ray shader.

```
//---Standard Automatic Parameters

float4×4 Wvp: WORLDVIEWPROJECTION;

float4×4 View: VIEW;

float4×4 World: WORLD;

texture tex: TEXTURE;
```

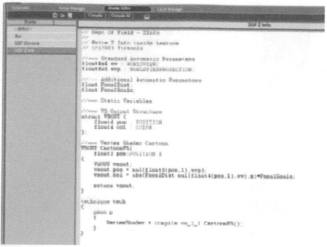

The Shader Editor lets you combine vertex shaders and pixel shaders. Press Compile to create a new shader available in the right column.

```
//---Manual Parameters

float density = 0.4;

float4 frontColor = {1, 1, 1, 1};

float4 backColor = {.5, .5, .5, .5};

//---Static Variables

//static float2 offset = 0.5-texelSize*0.5;

static float3 camDir = mul(World, View._m02_m12_m22);

static float4 minOpacity = {0,0,0, density*.5};

//---Vertex shader output

struct VS_OUTPUT

    float4 Pos: POSITION;

    float Opacity: COLOR;

        float2 TexCoord: TEXCOORD0;

//---Vertex shader function

VS_OUTPUT FrontVS(float3 Pos: POSITION,

        float3 Norm: NORMAL,

        float2 Tex0: TEXCOORD0)

    VS_OUTPUT Out;

//---Basic vertex transform

float4 screenPos = mul(float4(Pos, 1), Wvp);

Out.Pos = screenPos;

//---Compute reflection in UV0

float3 wNorm = mul(Norm,World);

Out.Opacity = density+dot(camDir, wNorm*density);

    Out.TexCoord = Tex0;

return Out;
```

```
VS_OUTPUT BackVS(float3 Pos: POSITION,

           float3 Norm: NORMAL,

           float2 Tex0: TEXCOORD0)

    VS_OUTPUT Out;

    //---Basic vertex transform

    float4 screenPos = mul(float4(Pos, 1), Wvp);

    Out.Pos = screenPos;

    //---Compute reflection in UV0

    float3 wNorm = mul(Norm,World);

    Out.Opacity = density+dot(-camDir, wNorm*density);

        Out.TexCoord = Tex0;

    return Out;

//---Textures Samplers

sampler texSampler = sampler_state

    texture = <tex>;

    Minfilter = LINEAR;

    Magfilter = LINEAR;

//---Pixel shader function

float4 FrontPS(VS_OUTPUT PSIn): COLOR

    float4 result = frontColor;

    float4 texCol = tex2D(texSampler,PSIn.TexCoord);

    float texAlpha = dot(texCol,1)/3;

    result[3] = clamp((PSIn.Opacity+texAlpha)/2.1,0,1);

    return result;

float4 BackPS(VS_OUTPUT PSIn): COLOR

    float4 result = backColor;
```

```
float4 texCol = tex2D(texSampler,PSIn.TexCoord);

float texAlpha = dot(texCol,1)/3;

result[3] = clamp((PSIn.Opacity+texAlpha)/2.1,0,1);

return result;

technique tech

    pass p

    {

        AlphaTestEnable = TRUE;

        AlphaFunc = GREATER;

        AlphaRef = 3;

        AlphaBlendEnable= TRUE;

        ZEnable = true;

        ZFunc = ALWAYS;

        ZWriteEnable = true;

        Texture[0] = <tex>;

        CullMode = CW;

        DestBlend = INVSRCALPHA;

        SrcBlend = SRCALPHA;

        VertexShader = compile vs_1_1 BackVS();

        PixelShader = compile ps_1_1 BackPS();

    }

    pass p2

    {

        CullMode = CCW;

        DestBlend = ONE;

        SrcBlend = SRCALPHA;
```

VertexShader = compile vs_1_1 FrontVS();

PixelShader = compile ps_1_1 FrontPS();

}

Let's apply a coded shader in the Material Setup window.

Step 2. Go to the Material Setup window for the 3D model of your choice.

The following illustration shows a Material Setup window for material of the skin of the body called torso_bin2. You can see it on the Shader Editor button on top of the window. This will let you choose a shader available for this project. The Params Shader window, located in the Material Setup window, lets you select parameters manually. In this case parameters are colors, yellow and blue, and density.

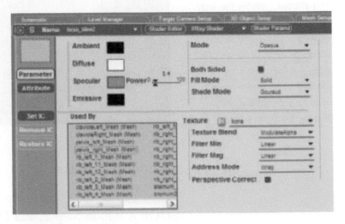

How to apply a coded shader in the Material Setup window.

4.2　Focus and Defocus

Goal: This tutorial shows how to experience the way our eyes choose to focus on objects. This shader imitates the way our vision chooses isolated objects within our line of vision or focal distance and does not focus on objects located in a circular area including peripheral vision.

Overview: Virtual cameras render crisp images of a virtual world regardless of the focal distance. Images from virtual worlds are too precise; they are lacking in unfocused peripheral vision or a circle of confusion that is less focused as you move away from the line of vision. Unfocused objects can be found in pictures of the physical world taken

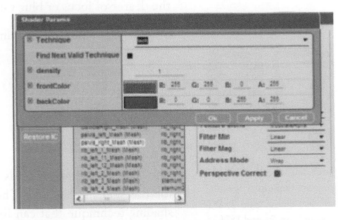

The Params Shader window, located in the Material Setup window, lets you select parameters manually. In this case parameters are colors, yellow and blue, and density.

through the lens of a camera. This setup combines three shaders to create a three-layer shading process. Open shader-defocus.cmo from the Interactive Textures folder on the companion CD-ROM.

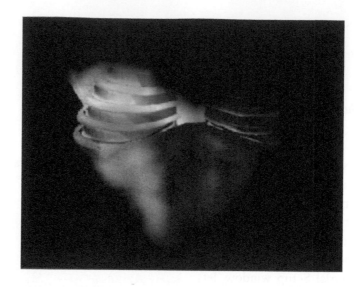

This shader imitates the way our vision chooses isolated objects within our line of vision or focal distance and does not focus on objects located in a circular area including peripheral vision. Shader by Zach Rosen.

Shading process:

Each step of the shading process requires a specific shader; Blur in Step 1, DOF Z Info in Step 2, and DOF Combine in Step 3.

Step 1. The image is processed and rendered to RGB. Information about the degree of focus or blur value is passed to the destination alpha-channel.

Step 2. The alpha-channel image is rendered.

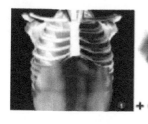 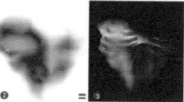

Step 3. The rendered image is composited with the alpha-channel to produce blurriness and transparency in the peripheral vision area.

Each shader is defined in the Shader Editor.

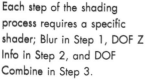

Each step of the shading process requires a specific shader; Blur in Step 1, DOF Z Info in Step 2, and DOF Combine in Step 3.

The DOF Combine shader is applied to the man_skin_Mesh material used for the skin of the body. The skin texture is replaced by an interactive shading technique that can follow the mouse pointer.

5   DISPLACEMENT MAPS

Goal: This tutorial shows how to create a terrain from a texture image. A flat mesh covered with a color texture can receive a topographic deformation according to a 2D image.

Each shader is defined in the Shader Editor.

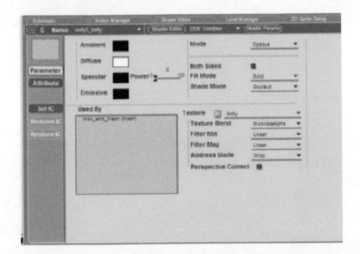

The DOF Combine shader is applied to the man_skin_Mesh material used for the skin of the body.

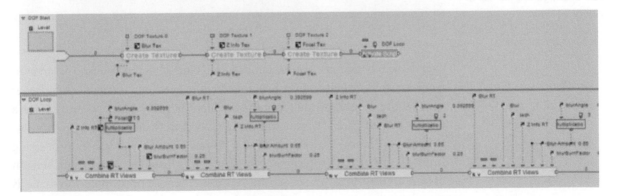

You can combine several shaders using the Combine RT Views building block.

The crater of the volcano is created from a displacement map, a black-and-white gradient image with a darker spot in the center.

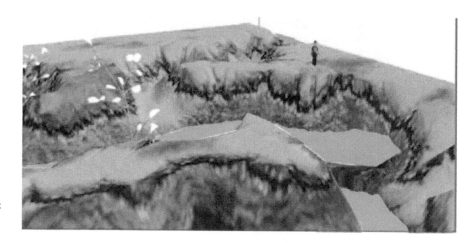

Displacement maps are very important tools to help illustrate and organize the flux of a story on a terrain.

Overview: In the previous example, the crater of the volcano is created from a displacement map, a black-and-white gradient image with a darker spot in the center. Modeling more details for the crater would require a more detailed mesh. If you add more polygons to the mesh, you increase the details for the volcano, but this comes at a price—longer processing time. The Texture Displacement building block uses a monochrome version of the texture image that can be red, blue, green, or alpha. Gray-scale is usually enough to create a topographic deformation of the mesh following the black-and-white gradient found in the texture image. The darker areas are pushed lower, and the brighter areas are pulled higher. Open texture-displacement.cmo from the Interactive Textures folder on the companion CD-ROM.

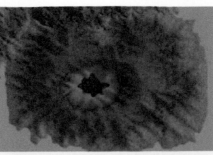

Left: Black-and-white image used as a displacement map. The darker areas are pushed lower, and the brighter areas are pulled higher. *Right:* Color image used as a texture map.

How to use it? Displacement maps are important tools to help organize the flux of a story on a terrain. This book covers several examples of terrains created for virtual spaces:

- Low to highly detailed terrains can be generated in Virtools or in Director 3D. Although this method is far from being the fastest and most precise method, the polygon overhead remains high without adaptative geometry.

- Highly detailed terrains used to create complex models (for example, a cityscape) can be generated in Bryce 5.0. Bryce terrains can be exported with an adaptative mesh that maintains the integrity of the terrain but reduces the number of polygons in areas with less detail. For more information, go to the tutorial about the Futuristic City at the end of this chapter.

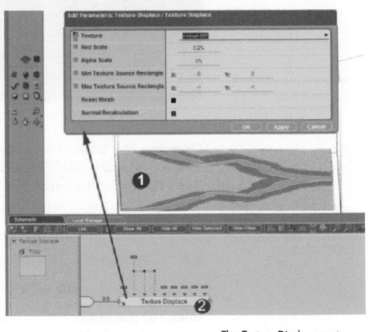

- Detailed terrains can be created in 3D modeling software and exported into Virtools or Director 3D.

The Texture Displacement building block uses a monochrome version of the texture image that can be red, blue, green, or alpha. Gray-scale is usually enough to create a topographic deformation of the mesh following the black-and-white gradient found in the texture image.

6   VERTEX PAINTING

Goal: The goal of this tutorial is to show how to create vertex painting combined with displacement maps to visualize and monitor large fields of data.

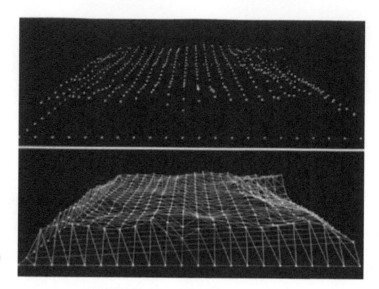

Vertex painting combined with displacement maps.

Overview: This tutorial shows how to render a texture on each vertex of a terrain. A texture called GreenRadial.jpg is applied to each vertex of the mesh. The size of the texture can be changed. Open vertex-painting.cmo from the Interactive Textures folder on the companion CD-ROM.

7   PROCEDURAL PAINTING

7.1   Painting in 3D with One Texture

Goal: The goal of this tutorial is to show how to paint 3D brush strokes on the surface of a 3D object.

Overview: This 3D painting technique borrows from several domains including cloning, virtual cameras, and texture mapping. Open procedural.cmo from the Interactive Textures folder on the companion CD-ROM.

When a mouse rollover is registered on the screen, the camera applies a brush stroke texture onto the 3D object, a sphere in this example. The brush stroke is applied directly under the mouse click. The 2D Picking building block registers the mouse position and sends a ray that finds the 3D object located under the ray. The intersection between the ray and the 3D object is used as a target for the camera. The Create Decal building block snaps a new polygon on the target with the brush stroke texture.

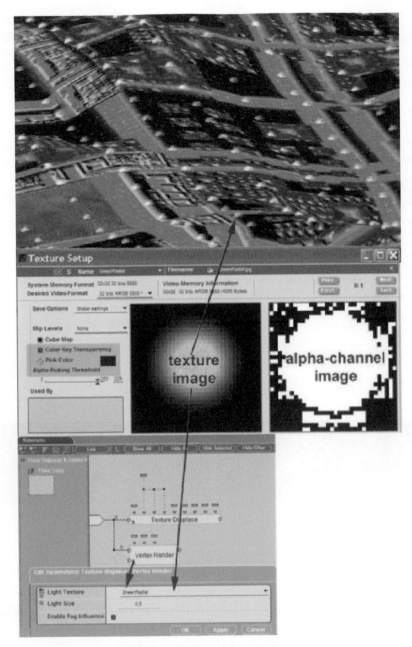

A texture called GreenRadial.jpg is applied to each vertex of the mesh. The size of the texture can be changed.

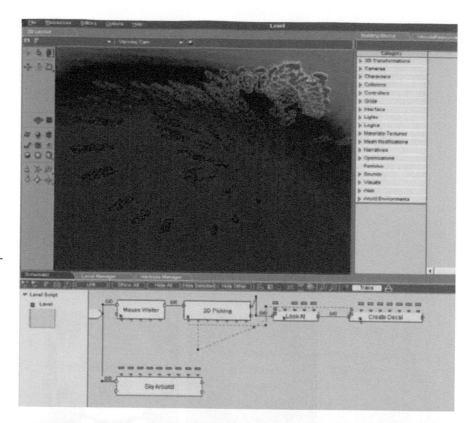

When a mouse rollover is registered on the screen, the 2D Picking building block registers the mouse position and sends a ray that finds the 3D object located under the ray. The Create Decal building block snaps a new polygon on the target with the brush stroke texture.

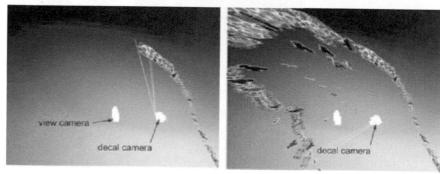

The target for the Decal camera follows the mouse pointer across the screen.

The brush stroke textures are made of bitmaps with alpha-channeled transparent edges. Clones of 3D polygons holding brush stroke textures are generated following the motion of the mouse pointer.

The same example can be updated with three brush strokes. The appearance of the brush strokes becomes unpredictable so as to recreate the physical feeling of painting with a real paintbrush. The Random Switch building block selects randomly from the list of brush strokes available.

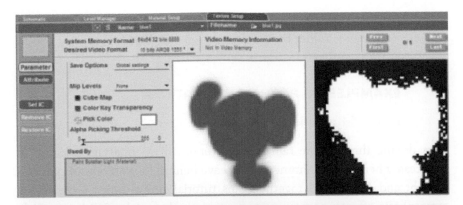

Example of an image with the black-and-white alpha-channel used as one of the brush strokes. The same example can be updated with three brush strokes. The Random Switch building block selects randomly from the list of brush strokes available.

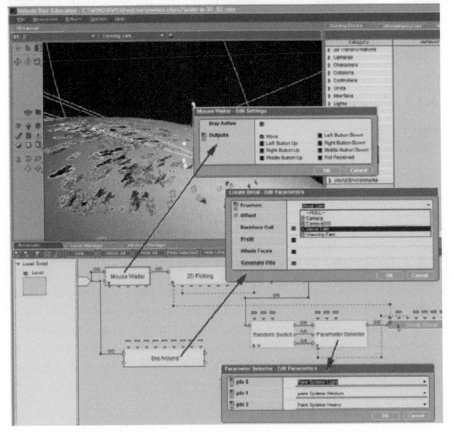

This example may be another way to recreate the physical experience of painting an object with a brush inside a virtual space. This can be a starting point for an interactive installation, blurring the borders between the real and the virtual.

How to use it? This example may be another way to recreate the physical experience of painting an object with a brush inside a virtual space. This can be a starting point for an interactive installation, blurring the borders between the real and the virtual. This last example closes our review of interactive texture mapping techniques. Compare several ways to implement 3D painting by opening 3DPaint1.cmo and 3DPaint2.cmo from the Interactive Textures folder on the companion CD-ROM.

8   EXAMPLES OF PARTICLE ANIMATION

8.1   Designing a Planar Particle Emitter

Cars driving through the rain in the avenues of Manhattan fade out in the steamy heat of the summer. They leave behind long streaks of particles of light and rain mixed together. This tutorial shows how to create the atmospheric effects in space using textures and particles. With a particle emitter, let's create a row of cars cruising through the Infinite City. Open infinite-city.cmo from the Interactive Textures folder on the companion CD-ROM.

8.1.1 Creating the Particle Emitter

Step 1. Create a new 3D frame, called "Emitter1." Scale the 3D frame 700 times on the Z axis. The 3D frame becomes a long narrow line.

Step 2. Add the Planar Emitter building block to the 3D frame. The parameters for the emitter are a flare texture and changes of color during the particle's life.

8.2   Designing Behaviors for Flying Particles

This tutorial shows how to set up a system of particles in motion. Particles fly inside a system where particle emitters create new particles and particle deflectors push them away. The particle emitters and deflectors constantly move on large paths. The particle emitter makes loops on the orange circle, and the particle deflectors move on the black circles. The deflectors push particles away from their initial trajectory.

Let's look at each element of the particle system.

8.2.1 Particle Emitters

Particle emitters are a type of vortex emitter that creates bursts of cloned particles. The particle generator moves on a circular path, and new particles are moving toward the center of the system. The particle emitter,

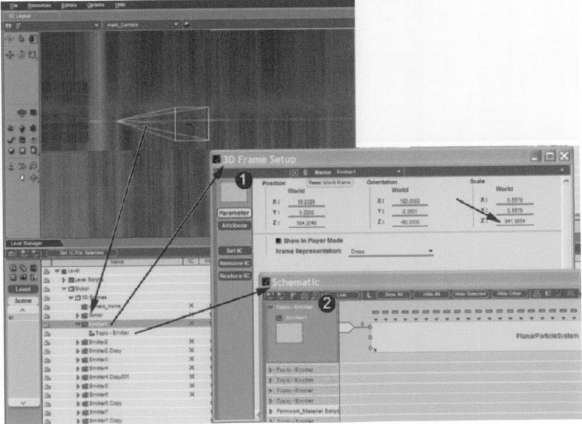

This illustration shows how to create a Planar Particle Emitter. Step 1 is to create a new 3D frame, called "Emitter1." Scale the 3D frame 700 times on the Z axis. Step 2 is to add the Planar Emitter building block to the 3D frame.

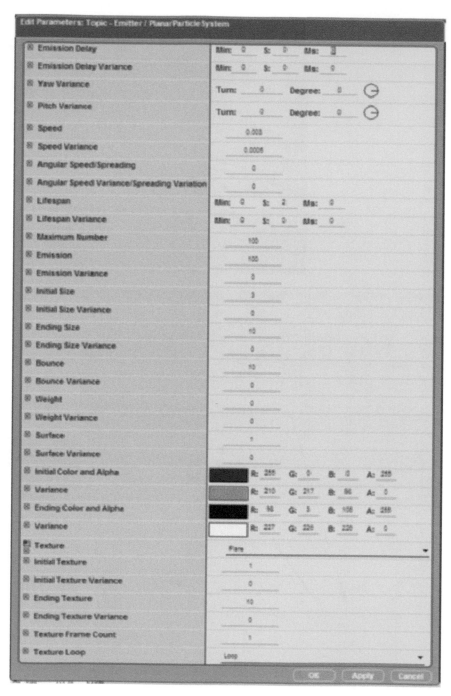

List of parameters for the Planar Particle Emitter including a flare texture and changing color during the particle's life.

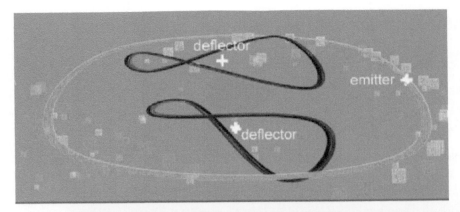

The particle emitters and deflectors constantly move on large paths. The particle emitter makes loops on the orange circle, and the particle deflectors move on the black circles. The deflectors push particles away from their initial trajectory.

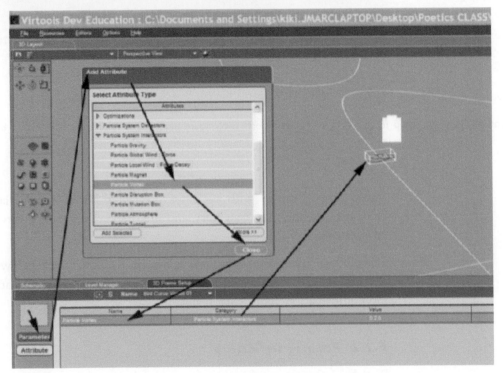

This illustration shows how to turn a 3D frame into a particle emitter. The vortex emitter can be found under Parameters > Add Attribute > Particle System Interactors.

called Particle Vortex, is an attribute of the 3D frame, shown as a white cross.

The illustration above shows how to turn a 3D frame into a particle emitter. The vortex emitter can be found under Parameters > Add Attribute > Particle System Interactors.

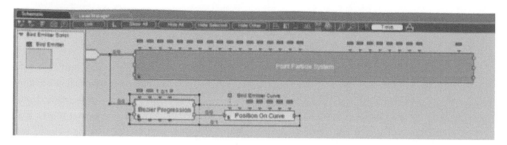

The 3D frame with the Point Particle System building block follows a large circular 2D curve, called Bird Emitter Curve.

This illustration shows the setup for the path of the deflectors. Each figure eight curve rotates on itself like a blender for particles.

The emitter is set up with a 2D texture called bubble.jpg. Variations of color, motion blur, size, trajectory, and life span are controlled through the emitter setup.

The 3D frame with the Point Particle System building block follows a large circular 2D curve called Bird Emitter Curve. Please go to Chapter 5 for more information about path behaviors.

8.2.2 Particle Deflectors

Particle deflectors push away new particles in many directions like a particle blender. The deflector is an attribute of the 3D frame found under Parameters > Add Attribute > Particle System Deflectors.

The 3D frame is set up on a path with a follow-path behavior. In this example, each 3D frame with a deflector attribute follows the path of a black figure eight curve. Each figure eight curve rotates on itself, helping to blend particles.

8.2.3 3D Objects Can Replace Textures in a Particle Animation Setup

3D objects (for example, small birds) can be emitted like particles. Birds can behave like the 2D textures shown in the previous example. The following tutorial shows how to set up a particle emitter with 3D objects. Please note that the particle emitter uses a Point Particle System building block.

Step 1. Go to Schematic, right click on the Point Particle System building block, and select Edit Settings. In the Point Particle System—Edit Settings window, select Particle Rendering > Object. Select OK.

Step 2. Right click on the Point Particle System building block. Select Edit Parameters. In the Edit Parameters window, select Texture > Bird. Select OK.

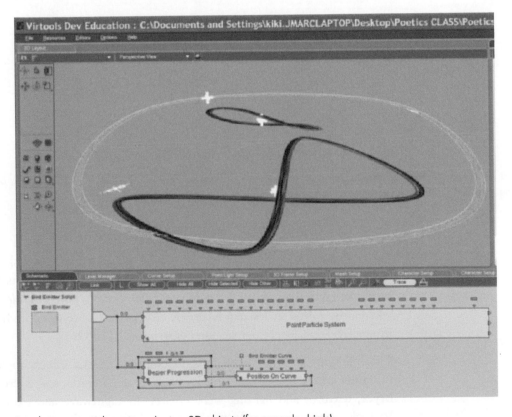

This illustration shows a particle emitter cloning 3D objects (for example, birds).

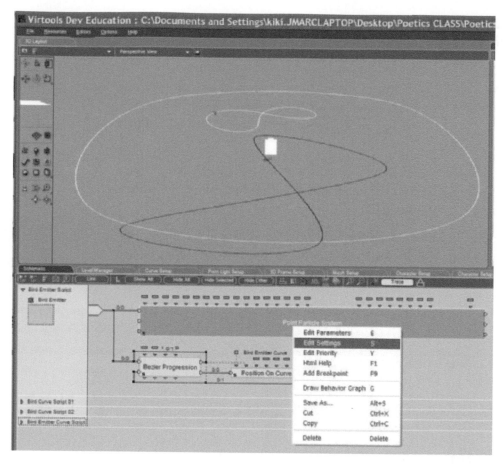

To set up a particle emitter with 3D objects, go to Schematic, right click on the Point Particle System building block, and select Edit Settings. In the Point Particle System—Edit Settings window, select Particle Rendering > Object.

9   DESIGN OF A FUTURISTIC CITY

Flying vehicles cross the futuristic cityscape of Los Angeles. The sky is illuminated by bursts of fire coming out of towers of chemical plants located inside the city. This opening scene of Ridley Scott's movie *Blade Runner* was an inspiration to design a futuristic city. Open future-city.cmo from the Interactive Textures folder on the companion CD-ROM.

The following tutorial shows how to combine several effects to recreate the atmosphere of the Los Angeles cityscape. The cityscape is created in Bryce; textures and lighting were created in Lightwave and exported to Virtools. Atmospheric effects are created around the streets and the buildings in Virtools.

This illustration shows the virtual cityscape created in this tutorial. Flying vehicles, generated by a particle animation system, roam above street level.

Several effects are animated together.

- Buildings are changing color and texture with time.

- Flying vehicles circulate above the city with unpredictable but organized patterns.

- Towers of fire illuminate the scene.

The effects presented in this tutorial include texture blending, a particle system with emitters and deflectors, and several particle emitters for the fires. The cityscape scene combines effects covered in previous sections of this chapter.

9.1   Designing the City

The design of a 3D model of the futuristic city can be automated by using a displacement map technique. Bryce, an automated terrain generator-program, is used in this tutorial.

9.1.1 Creating Textures in Photoshop®

Textures for buildings and streets at daytime and nighttime are created in Photoshop®. The textures use the Texture Blending behavior to simulate an accelerated 24-hour cycle. The following illustration shows examples of day and night building textures.

9.1.2 Modeling the Cityscape in Bryce

Bryce is a terrain generator using an adaptive mesh feature that lets you choose the number of polygons for your 3D model. Please note that this

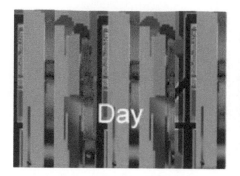

Textures for buildings and streets at daytime and night-time are created in Photoshop®.

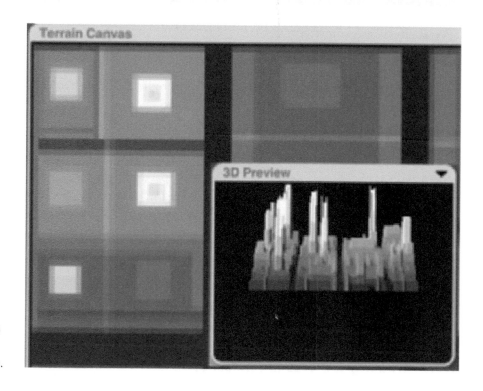

In Bryce, the Terrain Editor shows a preview of the cityscape, created with a 60 × 60 polygon square resolution holding 3,600 polygons.

feature is available only in version 5.0 and higher. For more information on Bryce, read *Real World Bryce 2* by Susan A. Kitchen, who will take you through a remarkable exploration of the software.

In Bryce, the Terrain Editor shows a preview of the cityscape, created with a 60 × 60 polygon square resolution holding 3,600 polygons. Terrains with higher resolutions allow the rendering of several blocks of the city with a fair amount of detail.

Please note that the terrain is rendered without color textures in Bryce because texture maps will be applied in Lightwave. When exporting the cityscape to Lightwave, a slider located at the bottom of the Export Terrain window lets you control the number of polygons for your model. The number of polygons for a cityscape alone can be around 15,000. The number of polygons goes down to between 5,000 and 10,000 for a cityscape with characters and vehicles.

9.1.3 *Textures, Lights, and Cameras*

The cityscape can be divided into several sections that can exported one by one. Special attention will be given to the street elevations seen at street level. Textures, lights, and cameras are created in the 3D software of your choice (for example, Lightwave). Export the scene to Virtools.

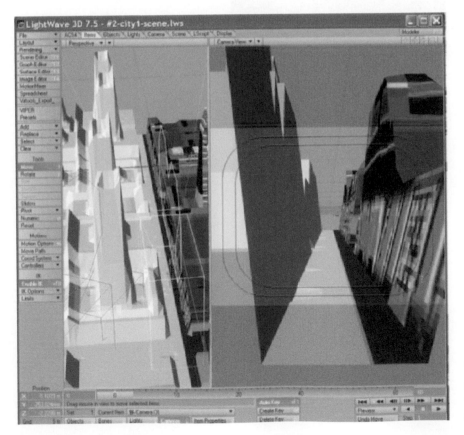

Textures, lights, and cameras are created in the 3D software of your choice (for example, Lightwave).

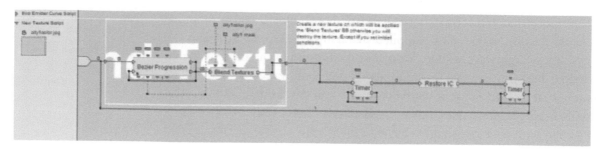

Buildings and streets of the city change appearance between night and day by blending a bright day texture with a dark night texture. The texture blending effect creates the illusion of a 24-hour cycle.

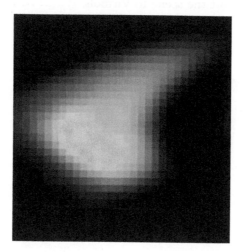

Example of 2D texture used for particle animation of the trail of the flying vehicles.

9.2   Particle Animation and Texture Blending in Virtools

9.2.1 Texture Blending

Buildings and streets of the city change appearance between night and day by blending a bright day texture with a dark night texture. A color texture is chosen for the day, and a gray-scale version of the same texture is chosen for the night. The texture blending effect creates the illusion of a 24-hour cycle.

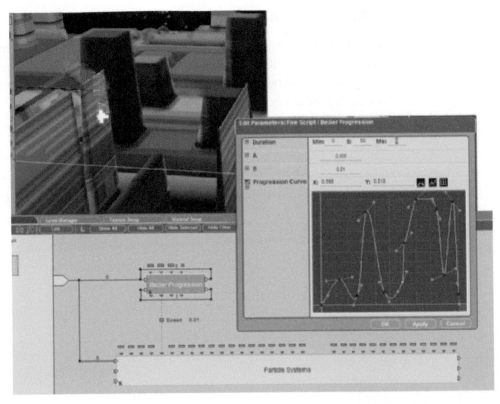

The bursts of fire coming out of towers burning natural gas in the middle of the the city are inspired by the opening scene of *Blade Runner* by Ridley Scott.

9.2.2 Particle Animation of the Traffic of Flying Vehicles

The particle animation gives the effect of fast flying vehicles. The complex particle animation system designed in the previous section is merged with the cityscape.

The image on page 152 is an example of 2D texture used for particle animation of the trail of the flying vehicles. The texture can be found at Virtools > Documentation > VirtoolsResources > Textures > Particles.

Example of 2D texture used for the fire.

9.2.3 Particle Animation of the Fires

The *Blade Runner* effects of running on top of tall buildings may be a reminder of the origins of a futuristic city associated with the fire coming out of oil rigs and tall derricks. The 3D frame with the fire emitter is placed on the top of the building.

The Bezier Progression building block can control the variation of intensity of the fire pulsations through a 30-second cycle.

An example of 2D texture used for the fire (fire.jpg) can be found at Virtools > Documentation > VirtoolsResources > Textures > Particles.

CHAPTER 4

Kinematics

1   INTRODUCTION

Kinematics, the fourth dimension of virtual spaces, goes beyond visual relationships between objects in motion. Object manipulations that take place in a dynamic space may involve obstacles, fixed objects, physics, hinges, and springs. Spatial relationships between moving objects can extend through space and time and affect many more levels than the sum of each moving object. For example, the motion of a ball depends on multiple parameters including elasticity of the ball, friction of the floor, and obstacles located on the trajectory of the ball. This chapter explains how to set objects in motion and how to create dynamic relationships between objects.

1.1   Kinematics

Kinematics shows how objects move and how the space around them is used during motion. We can look at the motion of objects as a process that encompasses space and the context of these objects in the virtual space. For example, an object rolling on a terrain can set other objects in motion and therefore change the topology of the space. The first section of this chapter covers simple manipulations in virtual space like moving an object, grouping several objects, reshaping, bending, and cloning. The second section focuses on complex manipulations of objects using several motions played in sequence. The third section about motion planning shows how autonomous agents can plan their movements before starting to move.

Designers need to know which motion techniques are more appealing to viewers. The perception of motion is very subjective. Viewers' eyes are quick to judge the qualities of the movement even before they see the details of the moving object. Viewers can intuitively break a simple motion into several subanimations. For example, viewers looking at the animation of a merry-go-round spinning inside a virtual garden can detect many components of the movement. They can notice when it moves quickly or slowly or if the center of rotation is slightly off. The eye can automatically detect something unusual and question the stability of the animation. The same goes for character animation. A small twitch in a character's walk can become so visible that the viewer may become frustrated with the movement of that virtual character.

1.2   Key-Frame Animations

Creating key-frame animations, a creative and precise process, starts with creating cycled animations inside 3D animation software. Canned animations, exported to a 3D authoring tool such as Virtools, are played on demand following the need of each character inside the virtual world. This process, described in detail in *Creating Interactive Actors and Their Worlds* by Jean-Marc Gauthier (Morgan Kaufmann Publishing), is one of the most widely used for character animation.

Although key-frame animations share similar starting and ending frames so as to loop seamlessly, the timing of a walk and the starting and ending points of a walk animation may change according to the character in the scene. For example, a character walking across a bar uses a walk cycle animation, a small animation that can repeat several times until the character reaches a seat. The same method is used to blend several animations together—for example, blending a walk animation and a sit animation. This solution may not work if several virtual characters walk inside a virtual bar and go to a different table every time they enter the bar or every time the viewer replays the scene. Character animations cannot be recorded in advance when a character changes path or actions all the time.

The last section of this chapter shows how to create motion planning by breaking down a character animation into small animations that can be assembled in a sequence. The result is a virtual actor planning a sequence of prerecorded animations while following a path.

Let's take the example of a character finding the best path for walking toward an empty chair. Try to sketch on paper a sequence of animations of a character walking to a chair and sitting in the chair. The main possibilities for the chair animations are as follows:

1 There are many ways to reach a chair: from the right, from the left, or from behind the chair. The placement of the character with his or

Before walking through the bar, each character needs to find an empty seat and plan the best path to reach this seat. Because the path of each character changes all the time, the character animation cannot be recorded in advance.

her back facing the chair is crucial before having the character sit in the chair.

2 A character can adjust the chair before sitting down—for example, dragging the chair away from the table.

3 Characters avoid bumping into each other when they stand up or are seated around the table.

4 A character can sit on all or half of the seat of the chair. A character can sit in the chair at an angle ranging from 0 to 180 degrees.

5 A character can move to slide under the table while seated on the chair.

Several simple animations can be created for each of these possibilities. Motion planning can help in the building of sequences made from simple animations. The number of sequences of animations possible for sitting in a chair is overwhelming.

What happens when you swap key-framed animations in the middle of a cycle or before reaching a transition key-frame? Canned animations with more than 3 seconds or 150 key-frames can be difficult to interrupt or blend on the fly with another animation without creating hiccups or twitching between animations. This problem will get more crucial when designing highly reactive 3D interactive content with a high demand for speed—for example, processing real-time facial animation according to someone's speaking or moving the fingers of a virtual hand holding a tool.

1.3   Why Use Event-Driven Motions?

We are interested in interactive systems that react to viewers in ways that cannot be prerecorded by the designer of a virtual space. The responses coming from the moving 3D objects or from the characters inside a virtual space need to be unpredictable but organized. Otherwise they will not sustain the viewer's interest. On the other end, viewers need to learn how to discover a virtual world and to influence it more than control it. Hybrid systems combining recorded animations and event-driven animations can react to the viewer's input in ways that create a delicate balance between the life of the virtual environment and the viewer's demands.

1.3.1 Hybrid Animation Systems

Dynamic control systems using artificial intelligence can mix canned animations and respond to the viewer's inputs. For example, artificial intelligence, using path-finding techniques, can decide how and when characters make choices about how to reach their goal. In the following illustration, virtual characters take advantage of space configurations to achieve their goal, shown with a red cross. Virtual characters follow hidden paths, shown with red lines in the illustration. The goals are assigned by the content management level system using artificial intelligence. After a while, viewers realize that characters moving along unpredictable paths never repeat the same pathway.

Dynamic control systems controlled by artificial intelligence can mix canned animations and at the same time respond to the viewer's inputs. Shown with a red cross in the illustration, virtual characters take advantage of space configurations to achieve their goal. Virtual characters follow hidden paths, shown with red lines in the illustration, that are assigned by artificial intelligence at the content management level.

1.3.2 *Parametric Animations*

Parametric animations are commonly used in the case of highly detailed motions—for example, changing expressions on a face or animating the fingers of a hand. The following illustration shows the importance of using parametric animations or event-driven animations to control the movement of the fingers of a hand. Each fingertip, attached to a 3D frame, can be moved like a string puppet. A chain of bones using inverse kinematics (IK) connects the tip of a finger to the palm.

The bones following each finger's movement can transform the polygonal mesh of the hand, resulting in the visual effect of a moving hand. Because the animation of a hand does not need to be key-framed, the full animation of the hand illustrated requires only five 3D frames moving along two directions. The animation of a 3D frame for each finger follows a simple set of parameters that consists of two axes of translation.

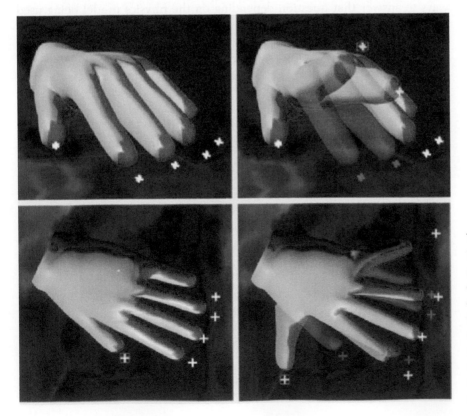

This illustration shows the importance of using parametric animations or event-driven animations to control the movement of fingers of a hand. Each finger can be moved like a string puppet using a 3D frame attached to the fingertip, shown with a white cross near the fingertip.

- The first axis running along the hand allows the 3D frames to extend and contract each finger.

- The second axis of translation running perpendicular to the fingers moves the 3D frame up and down. The same process would otherwise require multiple traditional key-framed animations for one finger. The Kinematics folder on the companion CD-ROM includes HandFK.cmo, an interactive example of the hand with forward kinematics. A step-by-step tutorial on setting up bones for a hand in Maya and Virtools is on the CD-ROM under Kinematics > Tutorials > skin.pdf.

Please note that the animation is not stored in active memory; event-driven animations are created on the fly. Parametric animations receive data flow from input devices or data-crunching applications.

1.4   Using Pseudophysics

What happens when two rigid bodies collide on a plane? There are many ways to answer that question depending on the physics model being used inside a virtual world.

Virtual worlds with pseudophysics use Newton's law about gravitation and collisions between objects. The following two Newtonian laws can be simulated in virtual spaces:

- *First law:* Every object moves in a uniform motion in a straight line or comes to a full stop if obstacles are applied on the object. For example, a 3D object with a translation transformation moves along a straight line until it hits an obstacle.

- *Second law:* The change of motion of an object is proportional to the force applied to the object. The following illustration shows a "driving" behavior used to move an object on a plane or on a terrain.

The driving behavior brings motion to a 3D object or character when the viewer presses keyboard keys to control moving forward, moving backward, and rotating clockwise and counterclockwise.

The Translate building block provides vector-based motion, which can be found under Building Blocks > 3D Transformations > Basic > Rotate or Translate. The Translate building block includes the following parameters:

vector coordinates and referential. Referential refers to where the vector force is applied. A referential can be a 3D object, character, or 3D frame.

The Rotate building block includes the following parameters: axis coordinates, angle of rotation, and the center of rotation of the referential or pivot point. The moving object can rotate around its pivot point or around the pivot point of another referential such as another 3D object, character, or 3D frame present in the virtual world. In the case of an object rotating around its own pivot point, the referential is the object itself.

The Keep on Floor building block makes sure that the object snaps on the floor below the object or that it remains at a constant distance above the floor.

When a moving object collides with an obstacle, the moving object stops in front of the obstacle or it can slide on the obstacle. The Object Slider building block can be found under Building Blocks > Collisions > 3D Entities > Object Slider. Please keep in mind that using the Object Slider building block requires creating a group of objects called "obstacles." The following illustration shows the driving behavior with an additional Object Slider building block for obstacles. The Kinematics folder on the companion CD-ROM includes Driving.cmo, an interactive example of driving a cube on a terrain with pseudo-physics. A step-by-step tutorial on setting up collision and floor detection in Virtools can be found in the CD-ROM under Kinematics > Tutorials > driving.pdf.

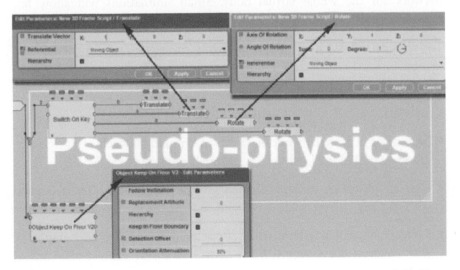

The "driving" behavior moves a 3D object or character when the viewer presses keyboard keys to control moving forward, moving backward, and rotating clockwise and counterclockwise.

1.5   Virtual Worlds with Physics

The physics approach looks at a balance of forces including friction, contact forces, gravitation, and other applied forces. The motion that results from using physics is an approximation of all related forces and their constraints. The physics engine chooses between forces to optimize the speed of the motion rendered on the screen. For example, the fast motion of a bouncing ball will be more convincing for the viewer than a slow but more accurate movement.

Let's look at a complex dynamic simulation of a virtual body where physics and dynamics play an important role. The project allows viewers to interact with a dynamic virtual person and to explore its physiologic system while the virtual person is moving in space and time. The virtual body needs to show reactions at a realistic pace without being slowed down by computing issues.

Each simulation builds a flexible network of 3D objects that can be connected together. For example, a simulation using the cardiovascular system may connect organs together though a dynamic system, whereas other organs connected though the respiratory system may not be dynamically involved. This process allows the addition of 3D objects to the physics system. In this same way, objects can also be removed from the physics system if they are not needed anymore. This allows the computer to run small-size dynamic simulations without the need to create a dynamic network for the whole virtual body. Small simulations will allow fast responses from the computer.

To "physicalize" 3D objects, the following properties are assigned to the objects:

- Fixed objects may only be used as obstacles like walls of the heart or arteries.

- Moving objects are dynamic elements of simulations, such as particle animation simulating blood flowing through arteries or valves.

- Convex objects are dynamic elements of the simulations often associated with moving objects, such as red cells flowing inside the bloodstream.

- Concave objects may be used as obstacles with irregular topography or with cavities—for example, heart cavities or inside tubular elements of arteries.

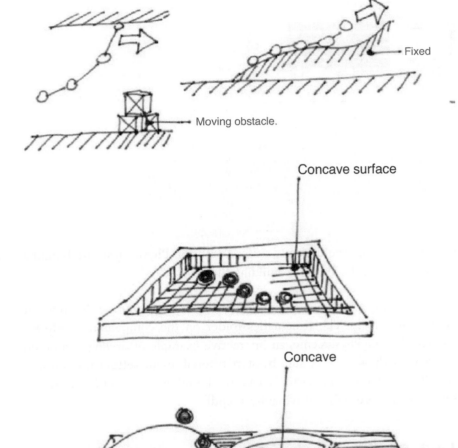

Concave surface

Concave

Moving objects are dynamic elements of simulations. Convex objects are dynamic elements of the simulations often associated with moving objects, such as red cells flowing Inside the bloodstream. Concave objects may be used as obstacles with irregular topography or with cavities—for example, heart cavities or inside tubular elements of arteries.

1.6 Moving Physicalized 3D Objects

Moving a Physicalized 3D object requires a different set of building blocks for simple translation and rotation.

1.6.1 Translation with Physics

The driving behavior is modified for a dynamic physics system using 3D objects with Physicalize properties. The viewer can control the translation

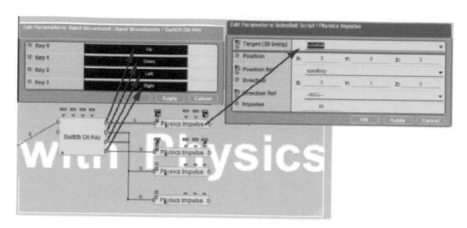

The driving behavior is modified for a dynamic physics system using 3D objects with Physicalize properties. The viewer can control the translations movements inside the dynamic system with the Physics Impulse building block.

movements inside the dynamic system with the Physics Impulse building blocks replacing the Translate building blocks.

Rotation inside a dynamic system with physics also requires a different set of building blocks. The Kinematics folder on the companion CD-ROM includes DrivingPhysics.cmo, an interactive example of driving a cube on a terrain with physics. A step-by-step tutorial about setting up collision and floor detection in Virtools can be found in the CD-ROM under Kinematics > Tutorials > drivingphysics.pdf.

1.6.2 Rotation with Physics

Viewers can control various rotation movements inside the dynamic system using the Set Physics Motor, Modify Physics Motor, and Destroy Physics Motor building blocks.

The following illustration shows a rotation behavior modified for a Physicalized object.

The Set Physics Motor building block, in Step 2, activates a dangling rotation with oscillations decreasing through time.

The Modify Physics Motor building block, in Step 3, creates a 360–degree rotation.

The Destroy Physics Motor building block, in Step 4, stops the rotation.

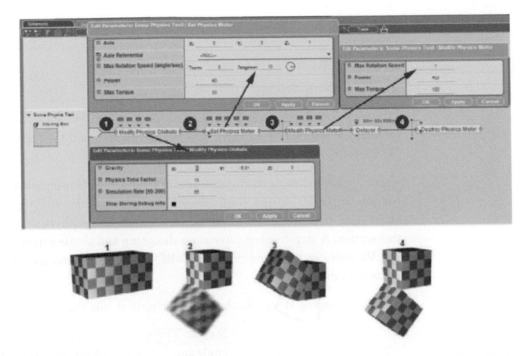

This illustration shows a rotation behavior modified for a Physicalized object. The Set Physics Motor building block, in Step 2, activates a dangling rotation with oscillations decreasing through time. The Modify Physics Motor building block, in Step 3, creates a 360-degree rotation. The Destroy Physics Motor building block, in Step 4, stops the rotation.

1.6.3 Adding New Objects to a Simulation with Physics

Action parameters set at the starting point of a simulation may change during the simulation. Dynamic systems often do not keep track of the actual state of a task being performed. To include new objects in a dynamic system, you will need to stop and restart the simulation. Please note that sometimes you may want to keep the virtual world playing back while changing the Physicalize properties of some 3D objects. This can be done by destroying the current simulation, removing the Physicalize properties of 3D objects, and recreating a new simulation. This process will ensure that new 3D objects are included in the simulation. The simulation can then run again with old and new Physicalized objects.

2　CREATING SIMPLE MOTIONS AND MANIPULATIONS

Creating simple motions such as falling, rolling, driving, colliding, and springing will help you to create complex dynamic systems with few

Physicalized 3D objects. I suggest playing the files that can be found on the CD-ROM to enjoy the visual richness of dynamic simulations using physics.

2.1   Rolling Stones

This section covers how to create a dynamic system of rolling stones falling on a terrain. The scene includes 3D objects of stones and the terrain. Each 3D object is Physicalized one by one.

There are several steps required to set up falling objects using physics. The Kinematics folder on the companion CD-ROM includes Rolling.cmo, an interactive example of stones falling on a terrain using physics covered in this section. A step-by-step tutorial on designing and cloning trees in Maya and Virtools can be found on the CD-ROM under Kinematics > tutorials > cloning.pdf.

Step 1. Dropping stones. Each stone falls until it hits the terrain.

Step 2. The stone hits the terrain at a certain angle and bounces back on the terrain. Please note that a flat floor is a fixed convex object and that a terrain with hills and valleys is considered a fixed concave object.

Step 3. The stone starts to roll and bounce freely on the terrain until it finds an obstacle.

In the following tutorial, we will set up stones that will start to roll and to bounce freely on the terrain until they find obstacles.

Let's set up the terrain with rolling stones.

Each rolling stone and the terrain are set up with Physicalize. In this case the rolling

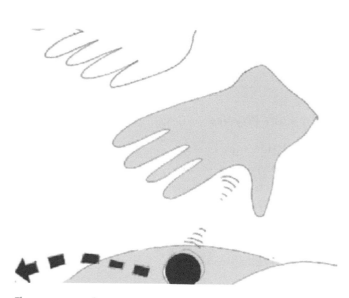

There are several steps required to set up falling objects using physics. The first step is dropping stones. Each stone falls until it hits the terrain. The stone hits the terrain at a certain angle and bounces back on the terrain.

stones and the terrain are highly detailed objects with irregular shapes and cavities that should be treated as concave objects. Please note that the simulation can also work with the rolling stones and terrain considered as convex objects, but the result will be different.

In the case of convex objects, the rolling stones will move faster because the physics system will consider them to be balls. The physics system will

Stones start to roll and bounce freely on the terrain until they find obstacles.

ignore the topography of the terrain and consider it a flat surface analogous to a bowling ball rolling on a completely flat and smooth wooden floor.

In the illustration on page 169, Step 1 shows how to Physicalize the terrain as a concave object. Step 2 shows that the terrain is a fixed object. Step 3 shows how to Physicalize one rolling stone as a moving object. Step 4 shows how to edit one rolling stone's properties as a concave object.

Please note that objects with Physicalize properties are set up as convex moving objects by default in Virtools.

This section showed how to create a dynamic system with few objects. The next section covers a dynamic system with hundreds of moving objects creating their own motion on the fly.

2.2   Multiple Collisions

This tutorial shows how to clone hundreds of moving objects, red balls, that are created by bursts of 50 new balls every 20 seconds in the center of the

Rolling stones on a terrain are illustrated in three steps. From top to bottom: view of the terrain before stones start to fall, stones hitting the ground, and stones rolling and bouncing on the ground.

terrain. You will also create 50 obstacles of white cubes. The result of this cloning process is a dynamic system where red balls bump into white cubes. The moving balls are creating their own motion in real-time on the surface of a terrain. The red balls fall and roll on the terrain until they hit the white cubes. Some balls bounce in another direction and some red balls accumulate along the obstacles, creating interesting patterns that can change through time. Please note that it is better to create the white cubes before the red balls to get the best results for collision detection. RollingBall.cmo, RollingBallsCubes.cmo, and RollingBallForest.cmo, from the Kinematics folder on the companion CD-ROM, are interactive examples of rolling balls colliding with obstacles covered in this section.

In the illustration on page 170, in Step 1 the white cube obstacles are cloned and the first burst of red balls spreads across the terrain. In Step 2, red balls follow the terrain and slide against the white cubes. In Step 3, red balls accumulate against white cubes, creating larger obstacles for incoming red balls.

Cloning 3D objects or characters and their properties is a powerful way to create interactive simulations. In this scenario, changing parameters such as friction from the floor, elasticity, and dampening of the red balls can directly influence the simulation. Visual parameters, like changing the size of the white cubes according to the number of collisions, can help keep track of areas where white cubes are in the way of red balls.

The outcome of this tutorial helps to illustrate the relationships between obstacles and moving objects through time. The simulation created in this tutorial is similar to the processes of erosion and sedimentation in a river.

During the erosion scenario, moving red balls push their way through an organized distribution of white cubes and rearrange the position of the cubes. The cubes may change color or size according to the number of hits received from the red balls. The new colored pattern shows how the erosion process takes place through time.

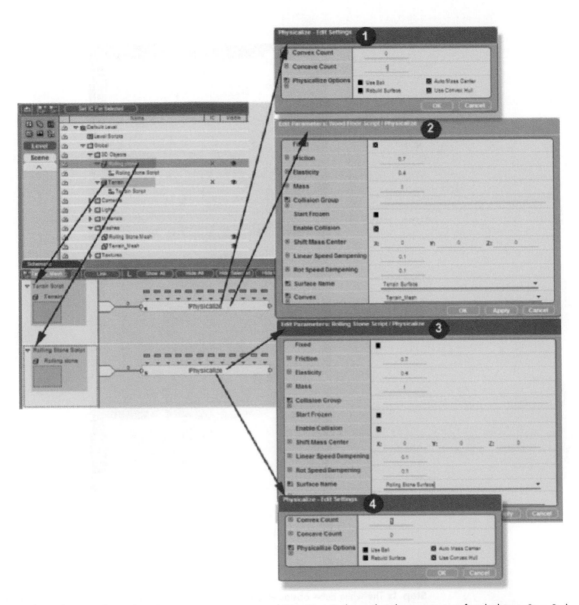

Step 1 shows how to Physicalize the terrain as a concave object. Step 2 shows that the terrain is a fixed object. Step 3 shows how to Physicalize one rolling stone as a moving object. Step 4 shows how to edit one rolling stone's properties as a concave object.

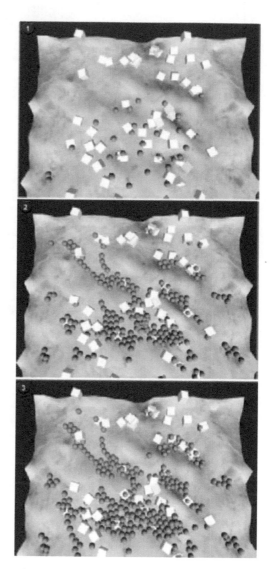

Step 1: The white cube obstacles are cloned and the first burst of red balls spreads across the terrain. **Step 2:** Red balls follow the terrain and slide against the white cubes. **Step 3:** Red balls accumulate against white cubes, creating larger obstacles for incoming red balls.

Moving balls push their way through an organized distribution of cubes and rearrange their position along the way.

Sedimentation can be illustrated in a similar way. For example, the sedimentation process in a river bed can be simulated by showing how floating objects, the red balls, are trapped by obstacles along the banks, the white cubes, and become themselves obstacles for incoming red balls. Number, color, size, and other parameters can change over time or in relationship to events taking place in the virtual world.

The following illustration shows a snapshot of a scene where clones of moving red balls collide with clones of fixed white cubes. The white cubes are grouped into a Fixed Objects group. Red balls are grouped into a Moving Objects group. Changing parameters inside the Fixed Objects behavior will affect all objects from this group and their clones.

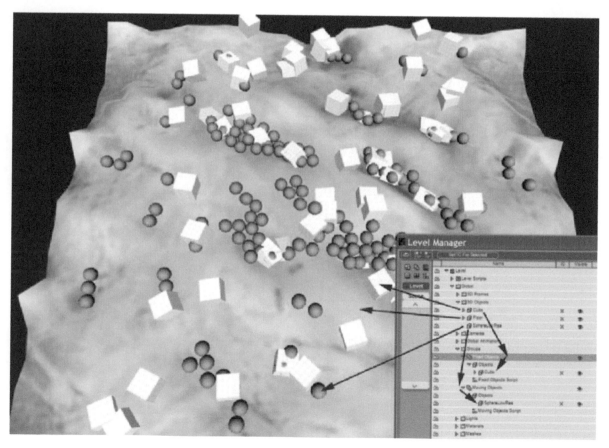

Snapshot of a scene where clones of moving red balls collide with clones of white cubes. The Fixed Objects group includes the white cubes. Red balls are grouped into the Moving Objects group.

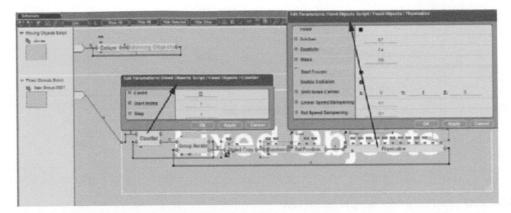

How to clone a group of 3D objects sharing the same Physicalize properties. The Counter building block controls the number of new clones. The Group Iterator building block makes sure that the behavior is assigned to each 3D object of the group. The Random and Set Position building blocks set a random distribution of 3D objects on the terrain.

The illustration above shows how to clone a group of 3D objects sharing the same Physicalize properties. The Counter building block controls the number of new clones. The Group Iterator building block makes sure that the behavior is assigned to each 3D object of the group. The Random and Set Position building blocks set a random distribution of 3D objects on the terrain.

2.3　Driving

2.3.1　Driving with Pseudophysics

This example shows how to drive 3D objects on a terrain using keyboard keys. The scene uses pseudophysics and gravity. Snake.cmo, from the Kinematics folder on the companion CD-ROM, is an interactive example of a snake crawling on a terrain covered in this section.

2.3.2　Driving a String of Balls with Physics

Driving, or rather "taming," a 3D object or a group of 3D objects with physics is an exciting experience that never repeats itself. In the following example, four balls drop on a concave floor and follow each other while the viewer is controlling the motion of the first ball. FourBalls.cmo, from the Kinematics folder on the companion CD-ROM, is an interactive example of four balls falling inside a box covered in this section.

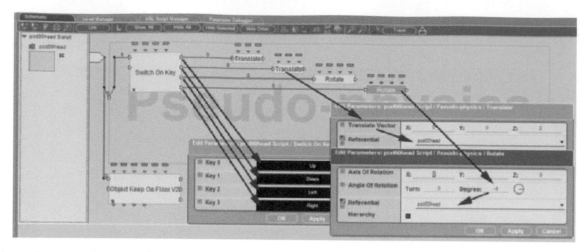

How to drive a 3D object on a floor using keyboard keys.

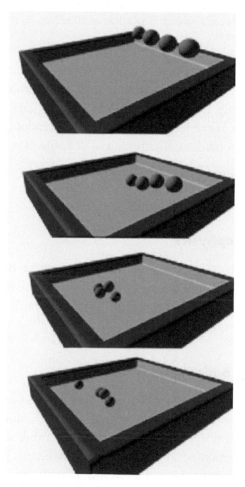

Four balls drop on a concave floor and follow each other while the viewer is controlling the motion of the first ball.

For the clarity of the tutorial, we pass the Physicalize property individually to each 3D object. Please note that other examples in this chapter show how to pass the same Physicalize property to all the members of a group of 3D objects. The red balls are set up with a motion controller to keep track of their motion changes while they move on the terrain.

Let's create parent–child relationships between red balls.

Steps 1 and 2 show how to use the Hierarchy Manager to set up parent–child relationships.

- Ball4 is the child of Ball3.

- Ball3 is the child of Ball2.

- Ball2 is the child of Ball1.

The Keep at Constant Distance building block keeps track of each ball in relationship with the previous one. Step 3 shows the new script added to each ball in the Level Manager. Step 4 shows how to set up the parameters for the Keep at Constant Distance building block for each ball in Schematics.

The order of the string of balls becomes as follows:

Ball4 keeps a constant distance from Ball3 and is the child of Ball3.

Ball3 keeps a constant distance from Ball2 and is the child of Ball2.

Ball2 keeps a constant distance from Ball1 and is the child of Ball1.

The following illustration shows details of the Driving with Physics script assigned to the first ball. Viewers can translate the first ball using the Physics Impulse building block. The Switch on Key building block controls translations along the X, –X, Z, and –Z axis.

2.4   Springs

Springs play an important role in our every day reality; they are regulating elements that can absorb and give back energy. Parts of our body—for example, the heart—are suspended inside a spring system. Cars are mounted on a system of shock absorbers, also a type of spring system. The following tutorial shows how to set in motion a moving plane suspended with four springs. Springs.cmo, from the Kinematics folder on the companion CD-ROM, is the interactive example covered in this section.

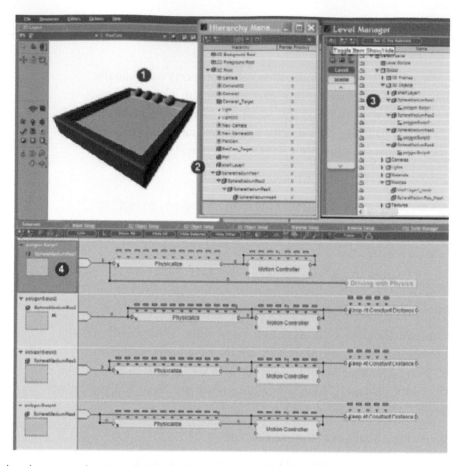

Steps 1 and 2 show how to use the Hierarchy Manager to set up parent–child relationships. Step 3 shows the Keep at Constant Distance building block added to each ball in the Level Manager. Step 4 shows how to set up the parameters for the Keep at Constant Distance building block.

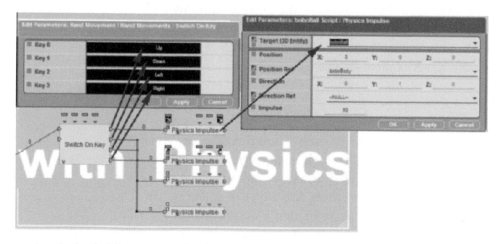

Viewers can translate the first ball by activating the Physics Impulse building block.

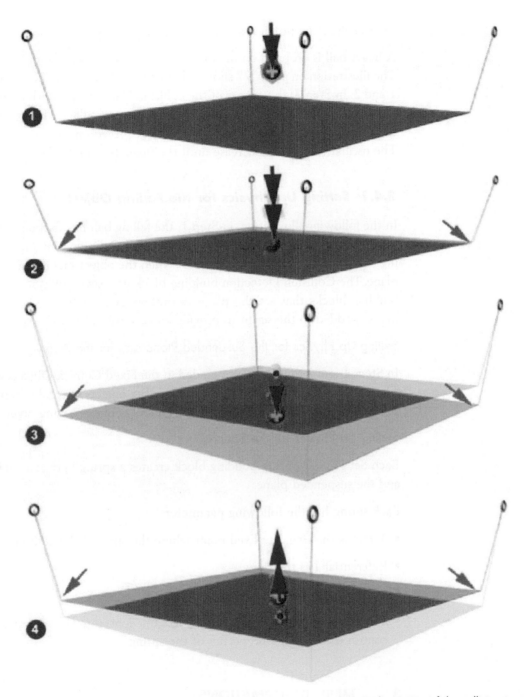

This illustration shows a heavy ball falling on the plane in Steps 1 and 2. In Step 3, the energy of the collision is sent to the springs. In Step 4, the spring system is set in motion and starts to oscillate. The oscillations of the springs cause the plane and 3D object placed on top of it to move.

Let's analyze the physics of a plane suspended with four springs.

A heavy ball is dropped on the plane and sets the spring system in motion. The illustration on page 177 shows a heavy ball falling on the plane in Steps 1 and 2. In Step 3, the energy of the collision is sent to the springs. In Step 4, the spring system is set in motion and starts to oscillate. The oscillations of the springs cause the plane and 3D object placed on top of it to move. The oscillations slowly decrease until the plane becomes idle.

2.4.1 Setting Up Physics for the Falling Object

In the following illustration, in Step 1, the falling ball has the properties of a Physicalized and convex moving object. The Collision Detection building block is added to the ball to detect when the object hits the suspended plane. The Collision Detection building block fires several Physics Impulse building blocks that set the plane in motion. A Switch on Key building block is added to the script to provide an easy way to test the setup.

Setting Up Physics for the Suspended Plane and for the Rings

In Steps 1 and 2 the rings are included in the Fixed Convex Objects group. The group receives a Physicalized fixed convex objects script. In Step 3 the plane has the properties of a Physicalized and convex moving object.

Setting Up Physics for the Springs

Each Set Physics Spring building block creates a spring between each ring and the suspended plane.

Each spring has the following parameters:

- Target is the ring, the fixed point where the spring is hooked up.

- Referential 1 is the ring.

- Referential 2 is the moving plane.

- The other parameters—including length, constant, and dampening—control the dynamic properties of the spring.

3   MESH DEFORMATIONS

The motion of some vertices from a mesh can control mesh deformations through time without affecting the whole mesh at once. Mesh

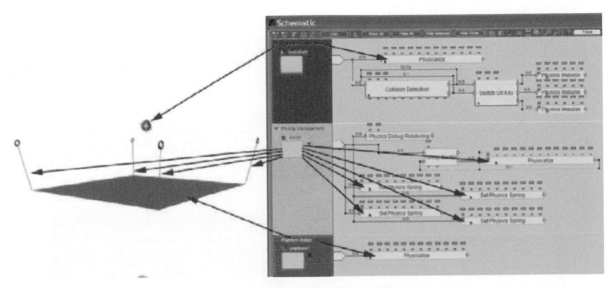

Setting up Physics for the falling object, the suspended plane, and the rings. Step 1 shows the falling ball with the properties of a Physicalized and convex moving object. The Collision Detection building block is added to the ball to detect when the object hits the suspended plane. The Collision Detection building block fires several Physics Impulse building blocks that set the plane in motion. Step 2 shows the script for the rings that are included in the Fixed Convex Objects group. The group receives a Physicalized Fixed Convex Objects script. In Step 3, the plane has the properties of a Physicalized and convex moving object.

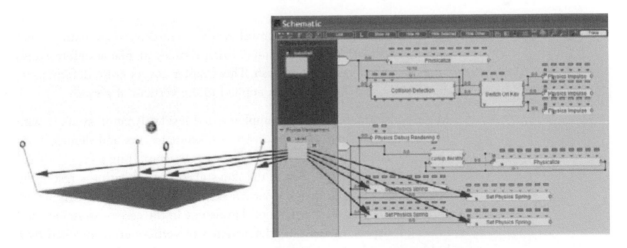

Each Set Physics Spring building block creates a spring between each ring and the suspended plane.

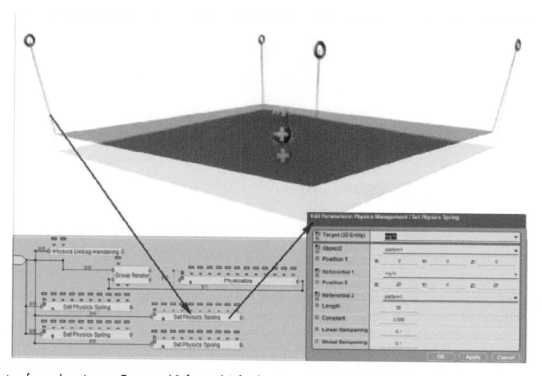

Parameters for each spring are Target and Referential 1 for the ring and Referential 2 for the moving plane.

deformations are easy to control with event-driven animations. For example, translating, bending, and twisting vertices are planar deformations applied to the vertices of a mesh. This chapter covers noise deformation, an example of 3D deformation applied to the vertices of a mesh.

Mesh transformations can be implemented inside dynamic systems with physics. For example, virtual surgery can simulate cuts and deformations through virtual skin created by a mesh. The surgeon lifting a flap of virtual skin can feel the "jelly" effect of the skin still connected with the rest of the virtual body. Vertices from the virtual skin can be constrained with hidden springs simulating the elasticity of skin tissues. In the case of stretching and moving a flap of virtual skin, deformation of vertices are controlled by a hidden network of springs connecting the skin with the bones.

3.1 Twisting and Bending

Twist.cmo and bend.cmo, from the Kinematics folder on the companion CD-ROM, are the interactive examples covered in this section.

3.1.1　*Twisting*

The twisting effect translates vertices along one axis—for example, the Y axis. Each vertex moves in relation to its corresponding index number inside the mesh structure.

In the following illustration, Step 1 shows that the vertex with index 0 is not moving. Step 2 shows the vertex of the tube with the highest index number moving with the longest translation. You can create the same twisting effect in other parts of the mesh by changing the Vertex Start Index parameter. Similar setups can be used for other kinds of mesh transformations such as taper, stretch, and explode.

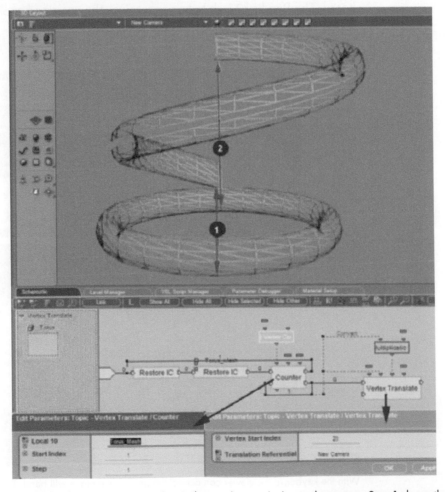

Each vertex moves in relationship to its corresponding index number inside the mesh structure. Step 1 shows that the vertex with rank 0 is not moving. Step 2 shows the vertex of the tube with the highest index number moving with the longest translation.

3.1.2 Bending

This example shows how to bend a plane when there are additional 3D objects placed on it. The following illustration shows red balls falling on the plane while the plane bends in different directions. The Keep on Floor behavior snaps red balls onto the surface of the plane. The red balls react to the various bending motions of the plane. If the plane bends down, the red balls fall off from the plane. If the plane bends up, the red balls fall inside the middle of the U shape of the plane. This virtual space with pseudophysics is only using the Keep on Floor building block.

The viewer can control the amount of bending of the plane with the keyboard. The Bending Mesh behavior receives the pressing key signal and triggers a linear progression of bending motions.

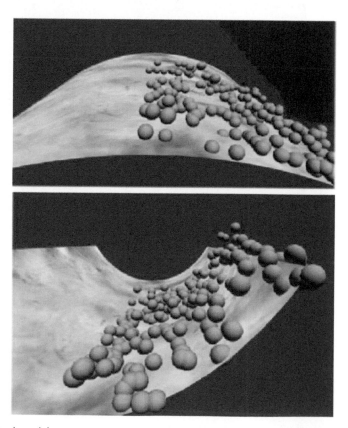

With the keyboard the viewer can control the amount the plane will bend. The red balls follow various bending motions of the plane. If the plane bends down, the red balls fall off the plane. If the plane bends up, the red balls fall inside the middle of the U shape of the plane.

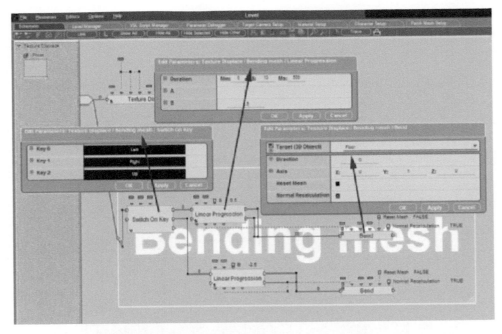

The Bending Mesh behavior converts the pressed key signal into a linear progression of bending motions.

3.2   Applying Noise to a Mesh

Noise is a 3D deformation creating spectacular morphing effects on 3D objects. The Noise building block applies a sine or wave deformation to the vertices of a mesh. The Noise building block can animate a transparent, blue mesh and turn it into waves at the surface of the sea.

3.3   Interactive Displacement of Vertices of a Mesh with a Live Video Image

Vertices from a polygonal mesh can be moved by a texture called a displacement map. More information about displacement maps can be found in Chapter 3. In this tutorial we use the image taken from a Webcam to animate the mesh. The live video texture from the Webcam is also used as the texture for the plane. VideoDisplace.cmo, from the Kinematics folder on the companion CD-ROM, is an interactive example of a displacement map using live video covered in this section.

The mesh processes the video texture in two different ways. In Step 1, a live video texture from a Webcam is applied as a texture on the material of the mesh.

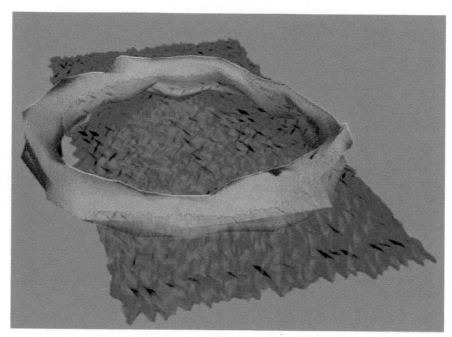

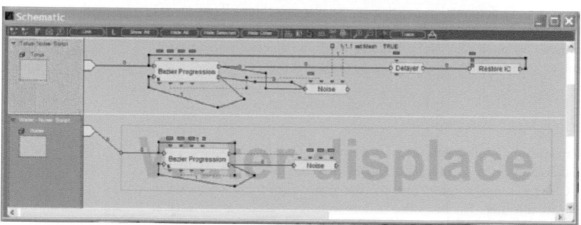

The Noise building block applies a sine or wave deformation to the vertices of a mesh. The Noise building block can animate a transparent, blue mesh and turn it into waves at the surface of the sea.

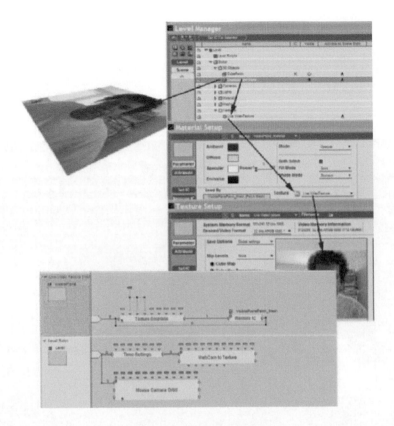

A live video texture is applied as a texture on the material of the mesh.

In Step 2, the video texture is acting as a displacement map, moving vertices aligned on a flat plane along the Y axis. The vertices move vertically according to the gray-scale and RGB pixel values of the video texture.

3.4   Moving One Vertex at a Time by Hand

This tutorial shows how to hand pick a vertex with the mouse and how to drag it away from the mesh. When you click on the mesh, a blue 3D cube marker snaps on the nearest vertex. The 3D cube moves with the selected vertex. The Move Vertex behavior can be combined with a spring system to simulate elastic fabrics, tissues, and virtual skin.

We are going to reuse and combine the simple movements and manipulations covered in this section for more complex moving structures. Let's look at ways to turn several red balls and tubes into the moving behavior of a snake.

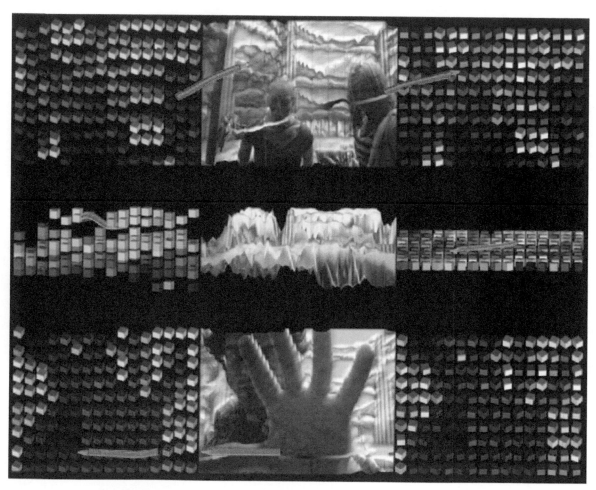

The live video texture is acting as a displacement map, moving vertices aligned on a flat plane along the Y axis.

Hand pick a vertex with the mouse and drag it away from the mesh. When you click on the mesh, a blue 3D cube marker snaps on the nearest vertex. The 3D cube moves with the selected vertex.

4   COMPLEX MOVING STRUCTURES

4.1   Snake

This tutorial shows how to create an animated snake that will respond to interactive controls. The head of the snake will move with translations and rotations. The tutorial shows how to build a modular structure that can mimic the motion of a snake on a terrain. The snake can be represented as a chain of joints connected with small tubes. All the objects in the chain follow each other during the motion. The moving body has a specific behavior coming from each joint following the footsteps of the previous one. A snake moves through a combined repetition of contractions and extensions of the muscles between the joints. The visual effect of the moving body is created by a slight attenuation of the length of the tubes connecting the joints. Snake.cmo, from the Kinematics folder on the companion CD-ROM, is an interactive example of a snake crawling on a terrain covered in this section.

This example shows how a chain of joints is scripted to emulate a snake behavior. The red balls follow each other while keeping a relatively constant distance with the previous module. Black tubular structures, with bones inserted between the red balls, can simulate the contractions and extensions of the snake's body. The tubular structures act as muscles attached to the red balls, constantly keeping track of the positions of the red balls.

4.1.1 Building the Snake

The following picture shows, on the left, the snake built with red balls, a green cone for the head, and black tubes inserted between the red balls for the body. On the right, the chassis of the snake is made of a chain of 3D frames. Creating a chassis of 3D frames connected together is easier to test and implement than creating the snake directly from 3D objects. If the chassis needs to be altered, other 3D objects can be added later. Creating a basic chassis lets you work on the snake behavior regardless of the context of a scene. You can reuse the moving chassis for other snakes, in other virtual worlds. Once the chassis is fully tested, various kinds of 3D objects, red balls, a green cone, and tubular springs, can be added to the snake. Red balls,

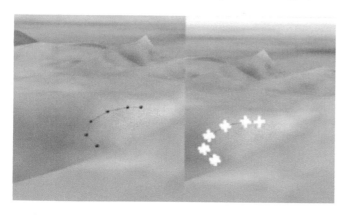

On the left view, the snake is built with red balls, a green cone for the head, and black tubes inserted between the red balls to create the body. On the right view, the chassis of the snake is made of a chain of 3D frames.

a green cone, and other 3D objects will be set up as children of the 3D frames.

Let's create the animated body of the snake.

Create one 3D frame above the terrain and rename it Pod00head. This 3D frame will be the head of the snake. Set Initial Conditions for the 3D frame.

Select Pod00head, and duplicate the 3D frame. Rename the new 3D frame Pod01 and move it along the X axis to create another section of the body. Repeat the same operation until you get at least five 3D frames for the snake. (The more times you repeat this process, the longer your snake will be.) Set Initial Conditions for all of the 3D frames so that the virtual world will remember the locations of the 3D frames.

The following illustration shows the lineup of 3D frames with new names. The 3D frames read as follows:

Pod00head—Pod01—Pod02—Pod03—Pod04—Pod05

Let's add behaviors to the joints of the body.

Go to Building Blocks > 3D Transformations > Constraints > Keep at Constant Distance.

The illustration on page 190 shows how to add the Keep at Constant Distance building block to each 3D frame.

- The Referential for the (n) 3D frame is the (n-1) 3D frame, the 3D frame in front of the current 3D frame.

- The Distance parameter sets the minimum distance between joints.

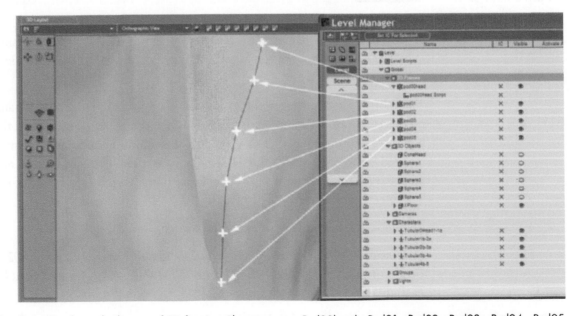

This illustration shows the line up of 3D frames with new names: Pod00head—Pod01—Pod02—Pod03—Pod04—Pod05.

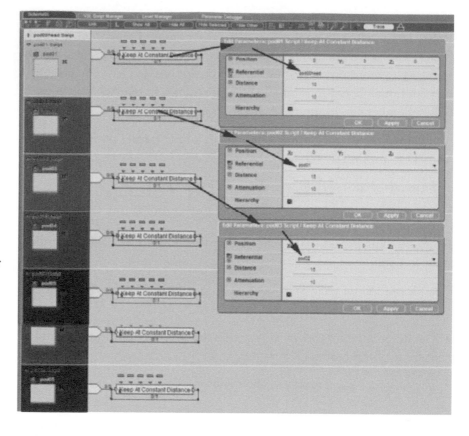

This illustration shows how to add the Keep at Constant Distance building block to each 3D frame. Referential for the (n) 3D frame is the (n-1) 3D frame, the 3D frame in front of the current 3D frame. The Distance parameter sets the minimum distance between joints. The Attenuation parameter sets a percentage to modulate the distance between two joints.

• The Attenuation parameter sets a percentage to modulate the distance between two joints. This creates the effect of contractions and extensions of the snake's body.

The tubular element inserted between 3D frames is modeled in a 3D animation software. Two bones are created inside each tube. The tube and the bones are exported together as a Virtools character. The tubular element is inserted between two 3D frames. The tubular element is duplicated and renamed for each segment of the snake. Set Initial Conditions for all of the 3D objects of the snake.

The updated snake structure is the following:

Pod00head—Tubular   head1-1a—Pod01—Tubular   head1b-2a—Pod02— Tubular head2b-3a—Pod03—Tubular head3b-4a—Pod04—Tubular head4b- 5—Pod05

Let's create "muscles" for the snake's body.

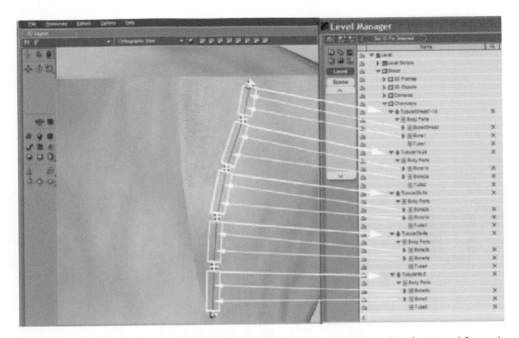

The tubular element is inserted between two 3D frames. The tubular element is duplicated and renamed for each segment of the snake.

The last step before moving the snake is to create a muscle behavior for the bones located inside each tubular element. Each bone is going to mimic the motions of the 3D frames located on each opposite end of the tube.

For example, let's look at one segment:

Pod01—Tubular1b-2a—Pod02

Tubular head1b-2a has two bones called *Bone1b* and *Bone 2a*. The segment can be rewritten as follows:

Pod01—Tubular1b—Bone1b—Bone 2a-2a—Pod02

The following illustration shows how to set up the Mimic building block for the bones. The Mimic building block is set up for each bone inside a tubular segment.

Go to Building Blocks > 3D Transformations > Constraints > Mimic.

- *Bone 1b* mimics the motion of the 3D frame, *Pod02*.
- *Bone 2a* mimics the motion of the 3D frame, *Pod01*.

Let's set up children and other 3D objects.

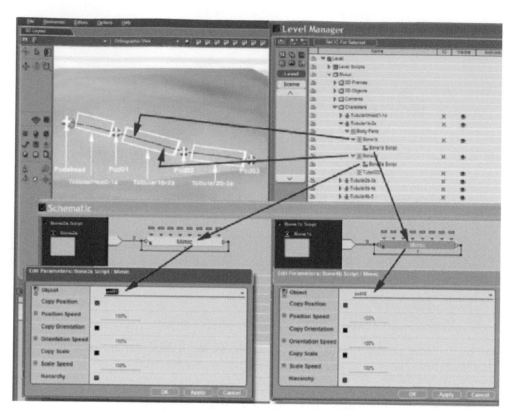

This illustration shows how to create muscles for the snake's body. The Mimic building block is set up for each bone inside a tubular segment.

Each 3D frame is the parent of a 3D object. Pod00head, the 3D frame for the head, is the parent of a spotlight and a camera in addition to 3D objects.

4.1.2 Driving the Snake

Pod00head is the 3D frame driving the snake. You can reuse the driving behavior for a virtual space with pseudophysics described earlier in this chapter. Please note that the Keep on Floor building block is used to keep the snake moving on the surface of the terrain.

5   INVERSE KINEMATICS

Birds flapping their wings or hands playing a musical instrument are examples of simple, fluid, and organic motions that require a perfect distribution of movement flowing from the main muscles and bones to the most

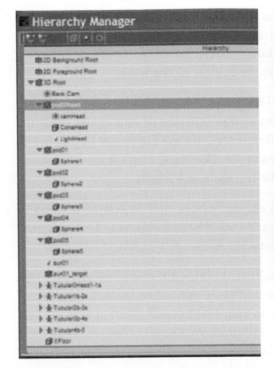

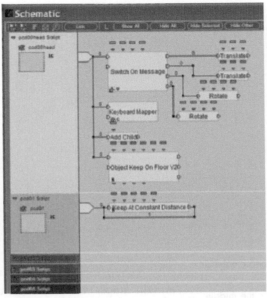

Script for Pod00head, the 3D frame driving the snake.

Each 3D frame is the parent of a 3D object. Pod00head, the 3D frame for the head, is the parent of a spotlight and a camera in addition to 3D objects.

delicate tips of the body. This kind of motion expresses the harmony between the movement applied to the whole body and a wide range of submotions applied to each joint. Animations using IK are easy to set up and a perfect solution for event-driven animation or hybrids between key-frame animations, path animations, and interactive animations.

5.1 Bird

This tutorial shows how to create an interactive animation using IK. The Kinematics folder on the companion CD-ROM includes Maya and Lightwave files for character animation covered in this section, and HandIK.cmo, an interactive example of a hand with inverse kinematics. A step-by-step tutorial on setting up an interactive hand with inverse kinematics in Virtools can be found in the CD-ROM under Kinematics > Tutorials > IK.pdf.

The IK chain of bones can control the motion of a single mesh without the need to key-frame every single bone. The chain follows a 3D object placed

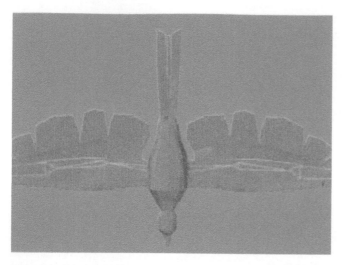

Top view of a bird created in Lightwave with two chains of bones running in the middle of the wings.

at the end of the chain. The following illustration shows a top view of a bird created in Lightwave with two chains of bones running in the middle of the wings.

The chains of bones can be tested in the 3D animation software of your choice using the Rotate tool. We use Lightwave Layout in this example. The visual result from the test will be the same in Virtools.

I suggest adding null objects at the end of each wing. The null objects will be converted into 3D frames in Virtools.

Although virtual birds can be animated without the need for weight maps or skin maps, you may want to add more control to the deformations of the wings. The weight map or skin map tool can add constraints to the vertices of the polygonal mesh for the wings. Weight maps help to control the resistance of vertices against the movement of the mesh. The resistance increases as you go away from the bone.

Weight maps or skin maps are bone dependent and cannot be created for a chain of bones. The animation of a heavy flight with slow-moving wings can be designed with a small area weight map located around the bone. A fast and nervous flight requires a larger weight map reaching more vertices around the bone area.

Please note that the tutorials in Chapter 2 and on the companion CD-ROM cover how to use weight maps or skin map tools in Maya. More in-depth information can be found in *Creating Interactive 3D Actors and their Worlds* by Jean-Marc Gauthier.

After importing the bird as a character, in Virtools go to Level Manager > Character > Bird > Body Parts. The following illustration shows the bones and mesh for the bird (showing up under Bird in the Level Manager). Rename the body parts and the 3D frames for each wing if necessary.

Let's set up the IK chains for the wings in Virtools.

We have imported a bird with one chain of bones for each wing. The IK Position building block creates the IK chain for each wing.

The chains of bones can be tested in Lightwave Layout using the Rotate tool. The visual result from testing in Lightwave will be the same in Virtools.

A heavy flight with a slow-moving mesh is provided with a small weight map, illustrated here, located around the bone. A fast and nervous flight requires a larger weight map reaching more vertices around the bone area.

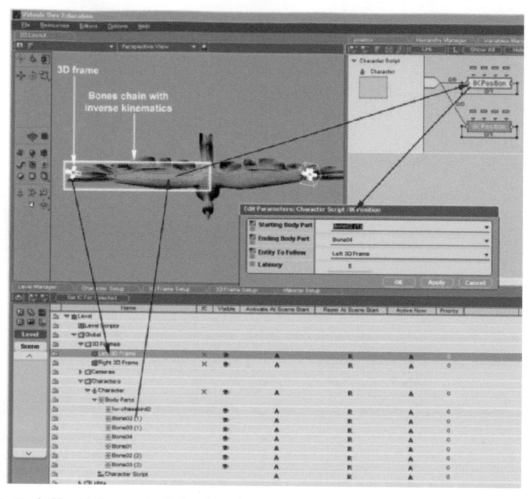

The IK Position building block creates the IK chain for each wing.

The parameters for the IK Position building block consist of the following:

- The Starting Body Part is the first bone of the chain located near the main body of the bird.

- The Ending Body Part is the last bone of the chain located near the tip of the wing of the bird.

- The Entity to Follow is the 3D frame located near the tip of the wing.

- Latency is the amount of delay in the movement between the end of the wing and the body of the bird. Latency helps to create the organic look and feel of the flight.

5.2   The Hand

This section shows how to rig a virtual hand for event-driven animations using IK and physics. A virtual hand holding a virtual tool is a fascinating metaphor for the manipulation of 3D objects in virtual spaces. VReam and other PC-based virtual reality applications created in the 1980s used a virtual hand as a metaphor of 3D navigation inside a virtual space. The virtual hand was designed as a visual relay for the mouse that could integrate traditional functions of the mouse pointer. The virtual hand can track positions of the mouse in space and bend each finger as a reaction to mouse clicks. The virtual hand is also rigged with physics to punch and grab objects. More information about setting up the virtual hand with input devices can be found in Chapter 8.

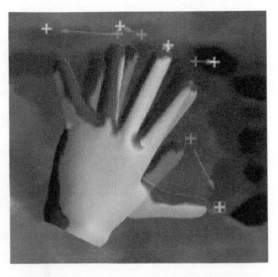

The virtual hand can track positions of the mouse in space and bend each finger as a reaction to mouse clicks. The virtual hand is also rigged with physics to punch and grab objects.

The following section shows how to set up the virtual hand with IK and physics. We use a single mesh polygonal hand with a bone structure following each finger created in a 3D animation software and exported to Virtools.

Moving the 3D frame vertically and horizontally controls the motion of each finger. Each chain of bones connects the palm of the hand and a 3D frame located at the tip of the bone. The motion of the 3D frame is distributed along the IK chain of bones running through each finger.

The setup in the virtual hand starts with the modeling and animation of the 3D hand in the 3D software of your choice. The following illustration shows a template for the chains of bones inserted inside the 3D hand. Renaming and indexing the bones for each finger will be a great help. I suggest that you create the null objects in the 3D animation software. The null objects will be interpreted as 3D frames once you export the hand into Virtools. After rigging the hand, the palm, and the fingers, you can test the bone structure by rotating each bone one by one. You can also create skin maps and weight maps for additional control of mesh deformations.

After importing the 3D hand as a character, in Virtools, go to Level Manager > Character > Hand > Body Parts. The bones and the mesh for the hand show up in the Level Manager.

Let's set up the 3D frames.

Go to Level Manager > 3D Frames. Check the list to make sure that each null object became a 3D frame for each finger. Rename each 3D frame with a finger's name if necessary. Each 3D frame moving a finger is a child of the hand. The fingers move with the rest of the hand. The illustration shows how each 3D frame representing each fingertip is also a child of the hand.

Let's set up an IK chain for each finger.

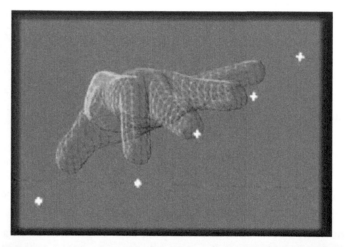

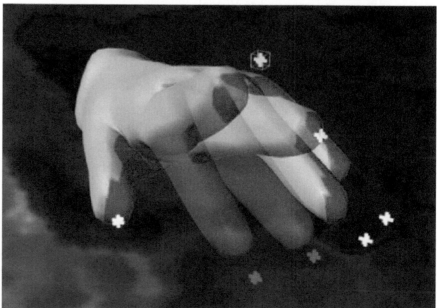

Moving 3D frames vertically and horizontally controls the motion of the polygonal mesh for each finger.

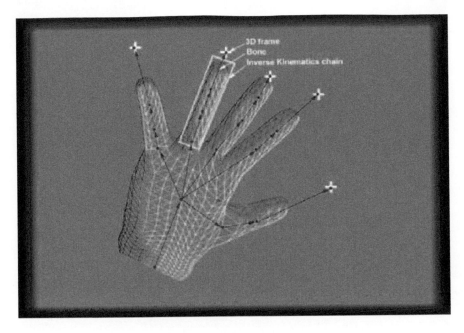

Each chain of bones connects the palm of the hand and a 3D frame located at the tip of the bone. The motion of the 3D frame is distributed along the inverse kinematic chain of bones running through each finger.

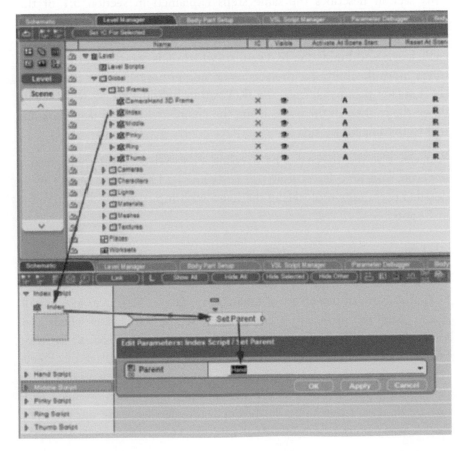

This illustration shows how to set up a 3D frame located at the fingertip as a child of the hand.

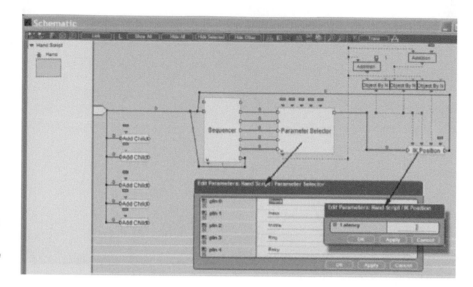

This behavior shows how to set up inverse kinematics (IK) for multiple chains of bones. The Sequencer building block goes through each 3D frame for each finger and builds the IK chain attached to it.

This section describes the same steps explained in section 5.1 of this chapter. The illustration above shows how to set up IK for multiple chains of bones. The Sequencer building block goes through each 3D frame for each finger and builds the IK chain to which it is attached. This method offers more flexibility for projects using numerous IK chains. Please note the Latency parameter from the IK Position building block, which slows down the motion of the bones along the chain, giving an organic look and feel to the finger's motion.

You can test the IK chains by pressing Play. Select a 3D frame near a fingertip, and drag it vertically or horizontally. Please note that a more detailed presentation of rigging the hand with input devices can be found in Chapter 8.

6   MOTION PLANNING

Spatial design can be the art of taking advantage of resources available in space. Using the surface of a table for crunching a nut is an example of improvisation with elements of space to create an instant tool. Only living creatures can think about this kind of ready-made assemblage of objects, which they may have learned from observing and perfecting what other people do. Observing people as they use their own hands reveals a wide culture of ways to set things in motion. How can we reuse our ability to invent new manipulations in virtual spaces?

Motion planning was invented for robotics applications where designers and builders needed to explore possible manipulations of objects in space and time. The following tutorial shows how to use several simple moving elements to plan a complex moving structure.

Motion planning is a way to project sequences of manipulations in time. We are constantly looking for the best sequence of manipulations to reach a goal with the minimum amount of energy expended. Virtual characters can use path planning, and other techniques inspired from research, on robotics to make choices before moving. Virtual characters explore several possible paths before choosing the optimum way to reach their goals. They can adapt their path according to incoming moving objects or to other characters. Viewers can endlessly explore a crowd of self-determined characters interacting together. Suddenly the idea of influencing preexisting movements replaces the traditional role of the viewer controlling the virtual space. The viewer becomes a tourist inside a virtual space where his or her presence is more or less acknowledged by other animated characters.

6.1   "3D Hand with a Pendulum" Project

The "3D Hand with a Pendulum" example explores ways to simulate a virtual touch and to manipulate tools with a virtual hand. The viewer can move the virtual hand above a terrain, creating visual associations between

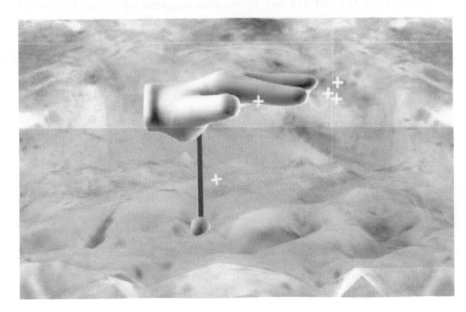

The "3D Hand with a Pendulum" project explores ways to simulate a virtual touch and to manipulate tools with a virtual hand.

the movement of the pendulum on the terrain and the feeling of touching the terrain. The pendulum is mounted on springs to convey the look and feel of friction on the surface of the terrain. The following illustration shows the 3D hand from the previous section with a pendulum suspended from the center of the hand. Pendulum.cmo, from the Kinematics folder on the companion CD-ROM, is an interactive example of a displacement map using live video covered in this section.

Motion planning requires the breakdown of a scene into several elements as indexed in the following illustration. The 3D objects in this scene are using physics.

- Step 1: The 3D hand is the centerpiece of the scene driving around the pendulum. The 3D hand uses the driving behavior with physics. The hand is above the terrain and placed inside the space created by four planar obstacles. A 3D frame controls each finger.

- Step 2: The spring system inserted inside the pendulum is similar to the snake created in Section 2.5 of this chapter. The spring is hooked on the hand and on the red ball. The top of the spring follows the hand. The bottom of the spring follows the red ball of the pendulum. The spring starts dragging the ball toward the hand. This happens when the elasticity of the spring equals the mass and friction of the red ball on the floor. Past a certain limit, the spring becomes overstretched and cannot move the red ball any more. This happens when the friction of the red ball with the terrain increases, such as the case of the red ball being stuck in a hole of the terrain. The string of the pendulum is made of a tubular structure with bones created for the snake in Section 4.1 of this chapter.

- Step 3: The red ball detects the presence of the floor.

- Step 4: The floor is a concave object with physics.

- Step 5: The red planes are obstacles to limit the motion of the 3D hand. These walls prevent viewers from driving the 3D hand outside the terrain.

- Step 6: 3D frames located at the 3D hand fingertips control the animation of the fingers.

The illustration on page 204 shows how to assign physics properties to a single object or to a group of 3D objects. The top view shows how to create

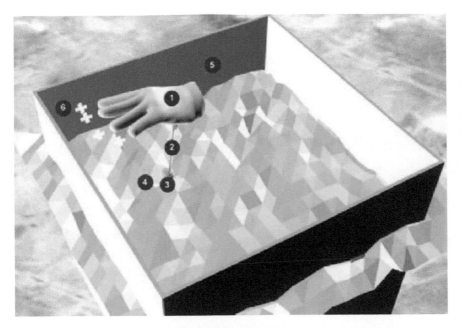

Step 1: The 3D hand is the centerpiece of the scene driving around the pendulum. Step 2: The spring system is inserted inside the pendulum. Step 3: The red ball detects the presence of the floor. Step 4: The floor is a concave object with physics. Step 5: The red planes are obstacles to limit the motion of the 3D hand. Step 6: 3D frames located at the 3D hand fingertips control the animation of the fingers.

a script for a convex moving object, the red ball. The bottom view shows how to create a script for a group of convex fixed objects, the obstacles. The Physicalize building block is assigned to each member of the group. The physics setup script for a group of objects goes through the list of 3D objects and characters contained in the group. Groups for this scene include fixed convex objects.

Let's set up the spring system.

The illustration on page 205 shows how to set up the Springs behaviors. The parameters for the Set Physics Spring building block are as follows:

- The target is the red ball.

- The first referential is the center of the tube of the pendulum.

- The second referential is the 3D hand.

- Dampening, constant length, and other parameters control the elasticity of the spring.

The physics setups for the scene are grouped together in one script called the physics management script.

Step 1 in the illustration on page 206 shows the physics setup for the 3D objects and characters:

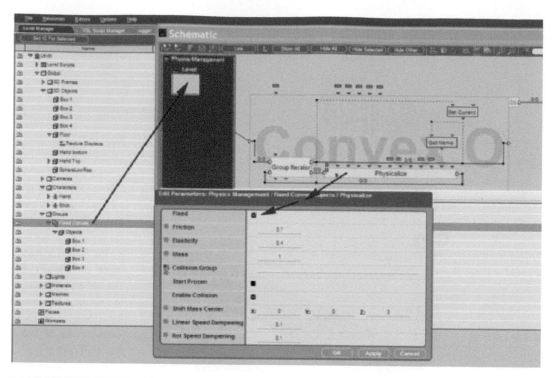

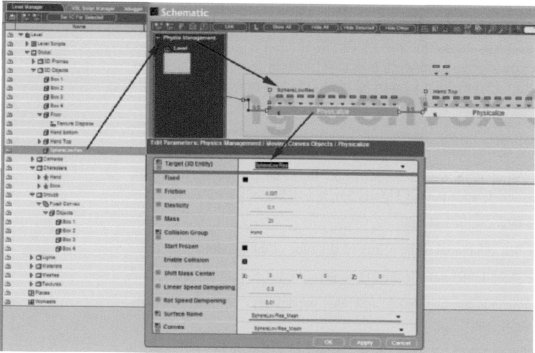

Top view shows how to create a script for a convex moving object, the red ball. Bottom view shows how to create a script for a group of convex fixed objects, the obstacles. The Physicalize building block is assigned the each member of the group.

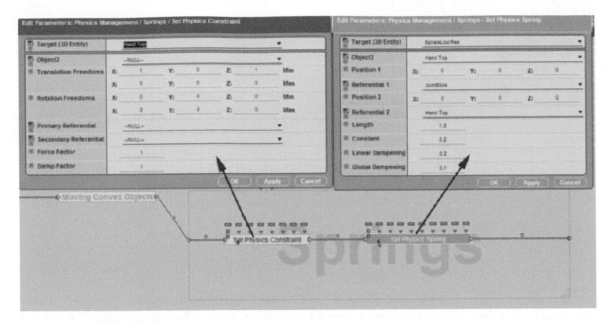

How to set up the Springs behaviors.

- Fixed convex objects include planar obstacles

- Moving convex objects include 3D hand, red ball, pendulum

- Springs for the pendulum

Step 2 shows the scripts for the terrain, a fixed concave object. The terrain uses a displacement map created in Chapter 3.

Step 3, in the following illustration, shows how the Hand Movement behavior is driving the hand with physics. The Switch on Key building block has two outputs controlling Physics Impulse in opposite directions.

To create additional outputs, right click on the Switch on Key building block and select Construct in the pull-down menu.

The Physics Debug Rendering building block comes in handy for showing which surfaces and lines of 3D objects are receiving physics properties. A material color selected in this example highlights the physics system in red.

6.2   Path Finding

This section shows how a virtual actor can plan and execute a sequence of animations while following a path. A self-determined character can reach a goal without using a predefined script.

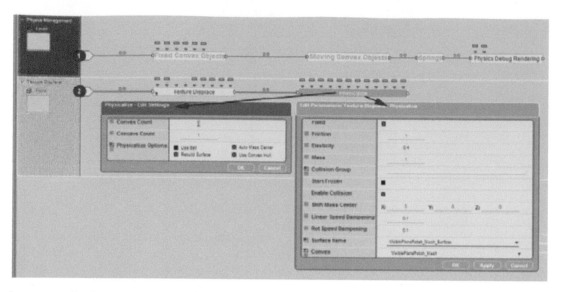

The physics setups for the scene are grouped together in one script called the physics management script. Step 1 shows the physics setup for 3D objects and characters. Step 2 shows the scripts for the terrain, a fixed concave object. The terrain uses a displacement map created in the previous chapter.

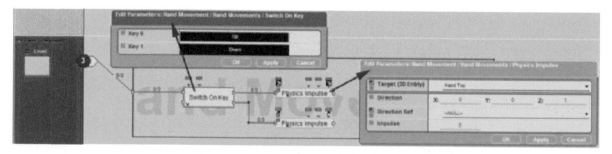

Step 3 shows how the Hand Movement behavior is driving the hand with physics.

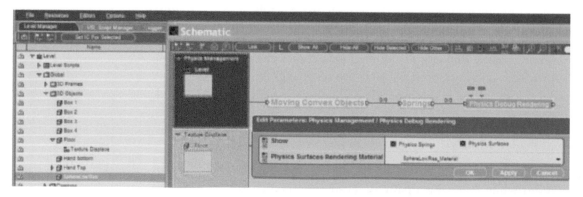

The Physics Debug Rendering building block shows which surfaces and lines of 3D objects are receiving physics properties.

Moving inside a physical space may include perceptions, reactions, and pre-defined knowledge that is grouped together inside a complex ecology of sensations. Several autonomous characters can interact the same way in a virtual space by sensing their proximity to each other and by sending messages to each other. These kinds of autonomous characters have the following properties:

- Autonomy: Virtual characters operating without human intervention, with their own control over actions in the virtual world.

- Communication: Various ways to acknowledge other characters or humans by some kind of language or message broadcasting system.

- Reactivity: Can sense the environment and respond in real time to changes.

- Self-determination: Goal-oriented virtual characters. They can make decisions to reach a target or to fulfill a goal.

6.2.1 Setting Up a Character with Path Finding on a Terrain

The following tutorial shows how to set up a grid with an obstacle layer and path finding. The tutorial is presented in two phases to help you; first you will test the collision detection system using the grid, and then you will add the path-finding features.

I suggest that in your Web browser you play back PathFinding.cmo, the scene from the Kinematics folder in the companion CD-ROM showing a character with path finding on a terrain. The demonstration works in two steps:

- Step 1. The viewer moves the goal, a white cube, inside the transparent areas of the grid.

- Step 2. The character finds the best way to reach the white cube while avoiding the obstacle areas, the yellow areas of the grid.

Setting Up a Character with a Grid and an Obstacle Layer

In this next example, the viewer controls the character animation. Create a new grid on top of the floor or on the terrain. In the grid setup window, you can create a new layer with a name and a color. In this example, the grid layer called Obstacles is yellow.

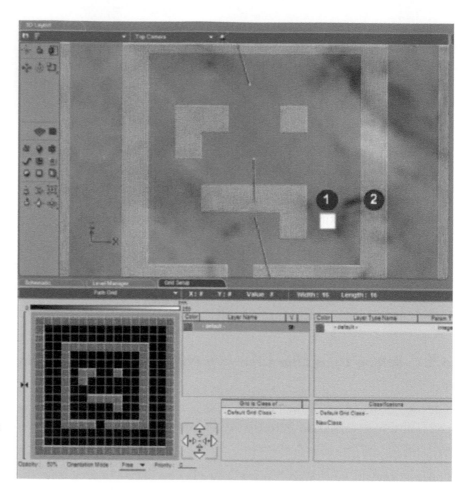

In the grid setup window, you can create a new layer with a name and a color. In this example, the grid layer called Obstacles is yellow.

The Layer Slider building block prevents the character from walking across the yellow layer called Obstacles. Please note the Influence Radius parameter that keeps the character at a certain distance from the layer.

Setting Up a Character with a Grid, an Obstacle Layer, and Path Finding

The scene requires a complex management system to combine interactions between the character, the obstacles, and the goal. During playback, the following levels work together to manage the character animation:

- Character walking toward the goal

- Collision detection with moving obstacles crossing the path of the character

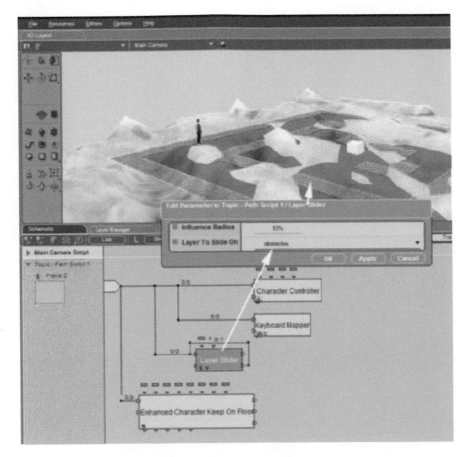

The character is controlled by the user. The Layer Slider building block prevents the character from walking across the yellow layer called Obstacles.

- Collision detection with other characters in the scene (for example, a scene can have several characters reaching the same goal)

6.2.2 Moving Goals through Time

The Aphrodisias project is a simulation of the life in the agora of an antique city. It includes self-determined characters walking across the marketplace of the antique city. The interactive characters create their own goals and plan the best path to reach their goals. As the location of characters' goals change during the day, characters decide what their priorities are. Some characters may decide not to follow their own goal. Aphrodisias.vmo, from the companion Kinematics folder on the companion CD-ROM, is the interactive example covered in this section.

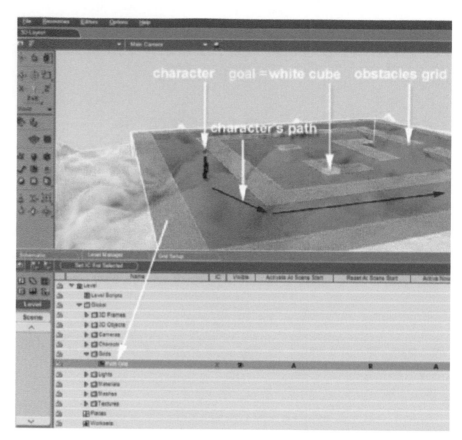

This illustration shows the levels working together to manage the character animation: character walking toward the goal, collision detection with moving obstacles crossing the path of the character, and collision detection with other characters in the scene.

The result is the look and feel of a fluid crowd that remains interesting for the viewer. The movements of the crowd are organized, unpredictable, and never repeat the same exact pattern.

The white crosses in the illustration on page 212 indicate the goals of the characters. You can follow the change of locations of the white crosses, the characters' goals, in the three snapshots taken during one day. Step 1 shows the morning, Step 2 shows the marketplace at noon, and Step 3 shows the marketplace in late afternoon. Please note that the characters cannot be seen in the top view because they are very small in relationship to the scale of the market place.

The virtual character automatically finds a path to reach its goal. Once the path is planned, the character starts walking toward its goal and avoids obstacles that may arise along the path.

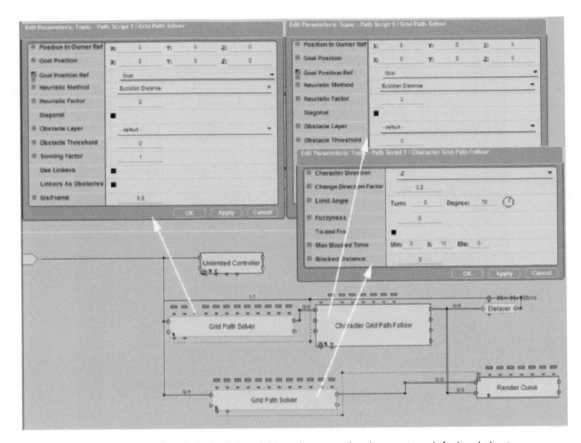

This illustration shows how to use the Obstacles layer of the grid to setup the character's path finding behavior.

The scene requires several levels of management of the interactions between characters and between characters and fixed objects. During playback, the following levels work together to manage the crowd of characters:

- Grid level defining fixed obstacles in the virtual world

- Global management of locations of character goals inside the virtual world

- Management of priorities for each character

- Decision-making process and path finding techniques for each character

- Collision detection of obstacles along the path of a character

- Collision detection between characters

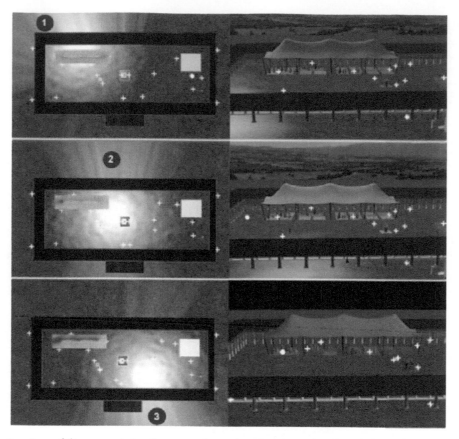

Top and perspective views of the agora during three snapshots taken during one day. Step 1 shows the morning, Step 2 shows the marketplace at noon, and Step 3 shows the marketplace in late afternoon.

1 Grid level with fixed obstacles. A green layer called Agora Obstacles is created in the grid. The green layer covers all 3D objects that are obstacles and that should be avoided by virtual characters. The obstacles include a fountain, the walls of a small temple, columns, a reflecting pool, the walls of the buildings, and the boundaries of the scene. Please note the border of the grid framed by a green line with interruptions. The green line represents the walls of the agora; the interruptions represent the doors where characters can walk in and out of the agora.

The green layer is traced on top of the floor plan or top view of the agora. Step 1 shows the area of the stores, Step 2 is the small temple, and Step 3 is the reflecting pool.

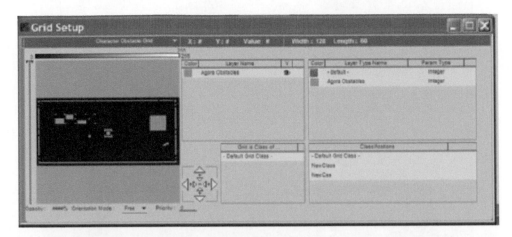

The green layer, called Agora Obstacles, covers all 3D objects that are obstacles and that should be avoided by virtual characters. The obstacles include a fountain, the walls of a small temple, columns, a reflecting pool, the walls of the buildings, and the boundaries of the scene.

2 Global management of locations of character goals inside the virtual world. The following illustration shows the location of white crosses, the characters' goals, in the space of the agora. You will notice that the white crosses are located outside of the obstacles (green areas).

Once the goals are defined each character has the freedom to edit its moves in relationship to its goal.

3 Management of priorities for each character. Characters create sequences of priorities that change with time. The list of possible priorities for a character includes the following:

• Waiting

• Walking randomly

• Exiting the agora (for example, in the evening characters exit the space)

• Going to a "hot spot" (for example, during the day the location of the hot spot may vary between the temple, the shops, or the reflecting pool)

• Characters can even choose not to reach their goal and to follow another character.

Ranking between priorities changes dynamically during the day. The Bezier curve changes the rank given to each priority during a 24-hour cycle.

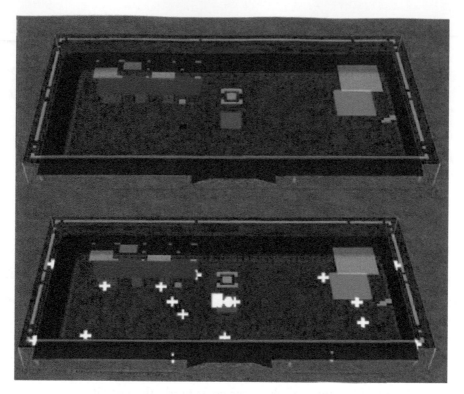

The top illustration shows the green layer traced on top of the floor plan, or top view, of the agora. The bottom illustration shows the location of white crosses, the characters' goals, inside the space of the agora.

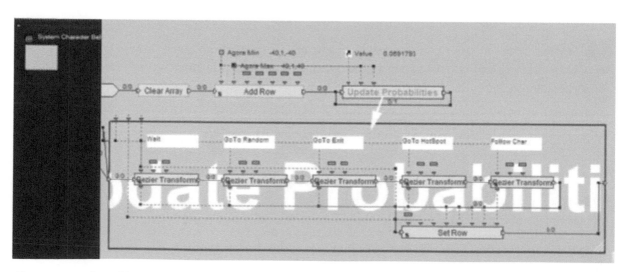

The sequence of possible animations for a character includes waiting, walking randomly, exiting the agora, or going to a hot spot. Characters can even choose not to reach their goal and to follow another character. Ranking between priorities changes dynamically during the day. The Bezier curve changes the rank given to each priority during a 24-hour cycle.

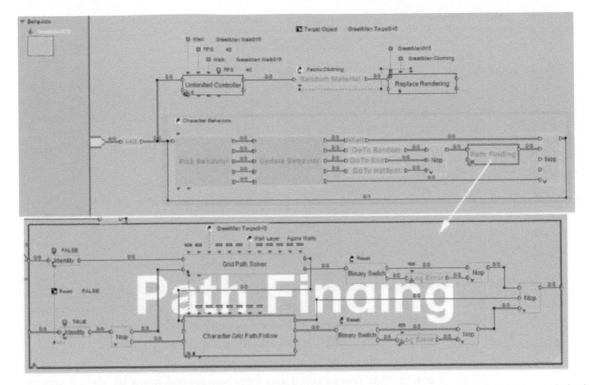

The decision-making process for each character includes picking a behavior and comparing the current behavior with the list of priorities. The outcome of this first round of choices will decide if the character activates the Path Finding behavior.

4 Decision-making process and path-finding technique for each character. The decision-making process for each character includes picking a behavior and comparing the current behavior with the previously noted list of priorities. The order of priority ranks differently according to the time of day a character updates its behavior. The outcome after this first round of selections will decide if the character activates the Path Finding behavior. The layout of the Path Finding behavior is similar to the one used in the previous section.

When moving characters and other moving objects (for example, horse carriages) cross a character's path, the character needs to go around the obstacle.

5 Collision detection of obstacles along the path of a character. In the case of a collision with moving objects, a character can use simple collision detection to slide on the obstacle and go back to its original path. In the case of a collision with another character, one character needs to learn how to give priority to another because both characters have the same urge to follow their own paths.

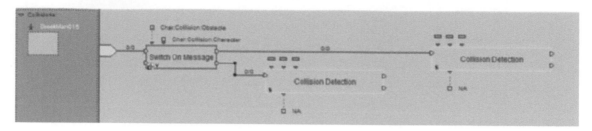

This script shows how a character updates its path when detecting 3D objects, obstacles, or other characters.

7   CONCLUSION

This chapter addresses how to create a wide range of dynamic systems of motion and how to control or influence them. The demand for simulation of lifelike systems and of natural organisms and the need to provide a better understanding of kinematics in the physical world creates a field of opportunities for kinematics in virtual worlds. This chapter covers aspects of kinematics that cross boundaries between several areas of virtual spaces. For example, movement and textures can be combined together to animate the surface of a texture and transform a 3D object over time. This chapter shows how to create dynamic systems with physics that can generate their own animations and self-determined moving objects and characters that can think and create a motion before using it. You will find similar strings of ideas in the next chapters about virtual cameras and advanced cameras.

CHAPTER 5

Interactive Paths

1   STORY OF A FAMOUS PATH

Power of Ten, the famous stop-motion animation created by film director and designer Charles Eames, illustrates the continuity of matter from the solar system to the atomic level. *Power of Ten* is structured around the idea of moving on a single trajectory to tell a story. Eames starts the story with shots across regions of outer space and then, in slow motion, takes us inside the atmospheric layers of planet Earth. The film ends up in the middle of a group of people having a picnic. After a pause at power zero, the camera resumes its vertiginous journey inside someone's hand. The camera stops when it reaches the atomic structure of molecules of the skin. After moving viewers from megapowers of ten at astronomic dimensions to the micropowers of ten at the subatomic level, Charles Eames informs us that he is going to rewind the course of the story being told. The camera moves backward along the path and again crosses an increasing number of powers of ten but this time at a very fast speed. This example by Charles Eames suggests the use of interactive paths to tell stories in virtual worlds. It also shows the possibility for viewers to control a journey in space and time.

1.1   Designing Paths

This chapter focuses on the design of curves or paths used to guide the motion of 3D objects, cameras, or characters inside a virtual space. Paths can be used indiscriminately by 3D objects, cameras, 3D frames, particles, or even interactive characters. Each path offers a controlled way to move through a virtual space. Paths are often invisible guides that allow the viewer to interact with the flow of a story in time and space. Paths are more than linear arrangements of events used to tell a story; they are similar to railroad tracks and switchyards with trains moving on them. The layout of paths across a virtual landscape or across a terrain can be compared with networks of railroad tracks, roads, and highways.

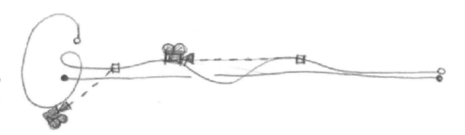

From top to bottom, sketches of a possible path for *Powers of Ten* going from megapowers of ten at astronomic dimensions, on the left, to the micropowers of ten at the subatomic level, on the right.

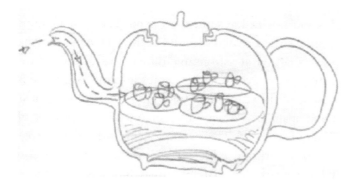

Paths are more than linear arrangements of events used to tell a story; they are also invisible guides that allow the viewer to interact with the flow of events of a story in time and space.

Although most paths are fixed trajectories in space and time, repeated travel along the same path does not always feels like a repetition of the same experience. Slight differences of lighting or changes of speed during each trip can change the viewer's experience. Arcade game players and viewers of virtual spaces often express that trips taken along the same path feel different every time. For example, tea poured from a teapot follows a path created by the design of the spout. How many times will the tea pour into the cup the same exact way after it leaves the lip of the pot? Although the tea follows the same path every time, I am intrigued by the fact that it seems impossible to repeat the exact same motion every time it follows the pouring path.

Another example of repetition and difference on the same path takes place inside the mine found in the movie *Indiana Jones and the Temple of Doom*. When shooting the scene, the crew built only a short, 15-foot long tunnel with miniature mine cars riding on railroad tracks. The crew shot along the same path, over and over again, with changes of lighting and various effects. The scenes seen in the movie are actually longer edited sequences from different versions of the same ride that have been pieced together. (Note from

George Lucas commentaries on the Indiana Jones DVD collection released in 2003 by Paramount Pictures.)

1.2   Managing Paths

Topics covered in this chapter include moving on paths, switching paths, and transitions between several paths. We will also look at ways to switch between being on a path and being set free without a path. The organization of the traffic of virtual objects moving along paths requires switches, transitions, and intersections that can be controlled from the interactive interface of the virtual world.

Two-dimensional paths are similar to a full-size road map opened on top of a landscape. We will see how they can be used for interactive characters automatically following the topography of a terrain.

Groups of acrobatic planes are a constant source of inspiration for the design of 3D paths, which can be very complex to reproduce in a virtual space. It is sometimes preferable to create a 3D path by recording and saving a path while moving a 3D object through a virtual space. The path can be

Two-dimensional paths are like a full-scale road map opened on top of a landscape. We will see how they can be used for interactive characters automatically following the topography of a terrain.

recorded in real-time when the 3D object is moving, then it can be saved and reused at a later date. In the following pages, we look at the example of recording a 3D path in the case of the flight of an insect.

A similar technique is helpful to train surgeons with virtual environments, like the human body, that would be otherwise difficult to explore with a mouse and keyboard. In virtual surgery, surgeons can move a sensor attached to a tool and record a 3D path inside a physical space. The information about the motion is passed to a virtual world where a virtual tool simulates the same motion inside a virtual space. The 3D path of the tool can be saved, and the actions can be replayed later for an evaluation of the surgeon's performance.

Moving up a tree, from the trunk of the tree to the smallest branches, can be achieved with one or several curves. Viewers can control speed and switch between paths when the tree is branching. This exploration of a branch can become increasingly complex when combined with other moving obstacles such as birds or insects or with wind.

In this chapter you will find the following:

1 How to set up the path of an insect flying along the branch of a tree

2 How to control the speed of an object on a path

3 How to create interactive motion control of an object on a path

Groups of acrobatic planes are a constant source of inspiration for the design of 3D paths.

4 How viewers can jump from one path to another

5 How viewers can move between several path cameras

6 How to switch a camera on the same path (for example, a camera can switch from path camera to a free camera)

7 How a moving object can detect obstacles on a path (for example, moving objects on a path can detect fixed obstacles and avoid them by changing paths)

8 How to record a path on the fly

9 How to create a racing camera inside a donut while following a 2D path

10 How to design particle animation for a futuristic cityscape inspired by *Blade Runner.*

Three-dimensional paths are more complex to create and can be recorded and saved while an object is moving through a virtual space.

This tutorial covers how to create paths with particle animation.

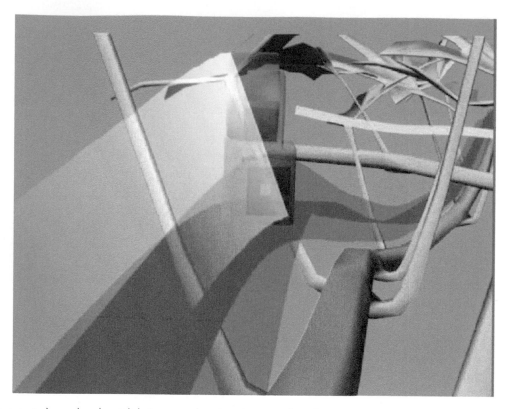

Viewers can control speed and switch between paths when the tree is branching.

Please note that in-depth information about free cameras, and especially free-range cameras such as the cocktail camera, can be found in the chapter about advanced cameras (Chapter 6).

2   PATHS TUTORIALS

2.1   How to Set Up the Path of an Insect Flying Along the Branch of a Tree

Goal: The goal of this tutorial is to learn how to build a path to move around 3D objects inside a tunnel. The Interactive Paths folder on the companion CD-ROM includes Path.cmo, the interactive example covered in this section. A step-by-step tutorial for building a branch in Maya can be found on the CD-ROM under Basic 3D Kit > Tutorials > branch.pdf.

Before and after creating the path.

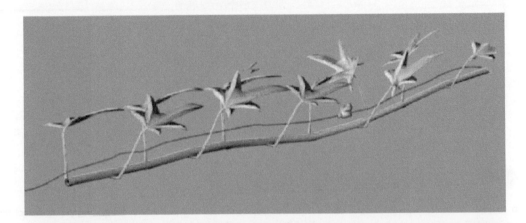

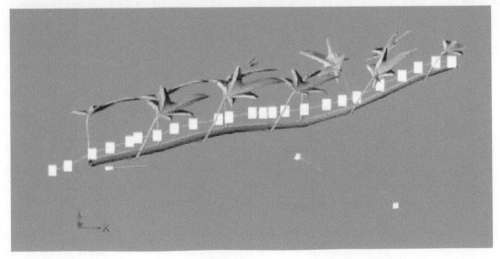

Top: Open path rendered during playback. Bottom: The same path with control points in white.

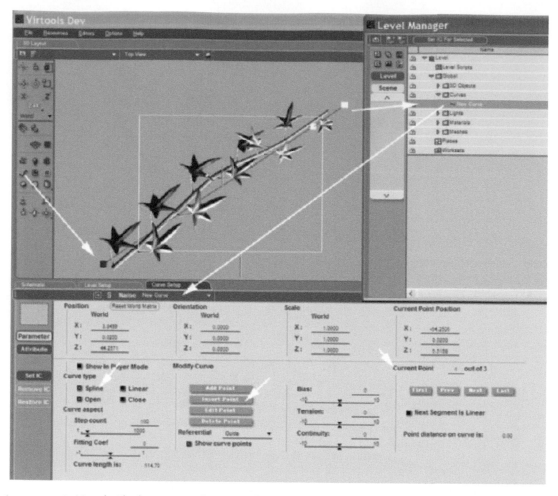

Creating a curve in Virtools. The best way to draw a path is to plot the starting point and the ending point. Select the insert point button in the Curve Setup window, and go to the 3D Layout window and click on the curve to add points.

A branch of a tree is a natural place to create an open path simulating the motion of ants and other insects walking up and down the branch. Insects can be followed by a virtual camera. This path is created in Virtools with the Create Curve tool. The best way to draw a path is to plot the starting point and the ending point. Select the insert point button in the Curve Setup window, go to the 3D Layout window, and click on the curve to add points. Go to a top view, and choose the Move tool along the XZ axis to adjust the position of the control points on the curve.

Go to a side view, and choose the Move tool along the Y axis to adjust the position of the control points on the curve. The same method can be used for a closed path. Let's add motion and interactivity to the scene.

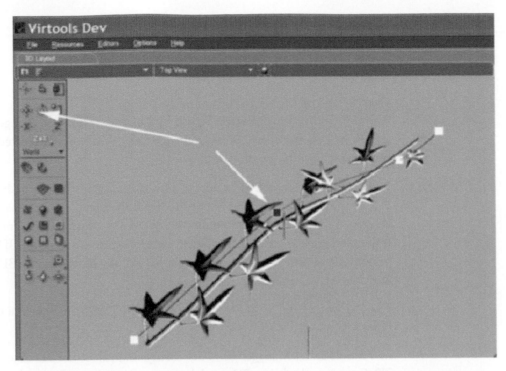

Go to a top view, and choose the Move tool along the XZ axis to adjust the position of the control points on the curve.

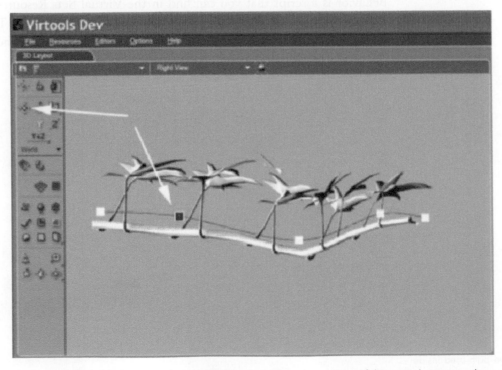

Go to a side view, and choose the Move tool along the Y axis to adjust the position of the control points on the curve.

2.2   Controlling the Speed of a Camera on a Path

Overview: This tutorial shows how to set up a 3D object on a path. In addition to following the path, the speed of the 3D object changes along the path. This tutorial requires you to import a 3D model created in the 3D modeling software of your choice.

In this example, a green 3D model of a movie camera follows a path created in Virtools. Although the speed of the 3D object can change over time, it is set by a Bezier Transform building block, which controls time, and a Position on Curve building block, which controls the position of the 3D object in function of time. The Position on Curve building block places a 3D object at a specific position on a 3D curve.

The following tutorial shows step by step how to set up a camera or a 3D object on a path.

In Virtools Level Manager go to 3D Entity, and select the 3D object called 3D Camera, which is a 3D model of a movie camera. Right click on the name and choose Create a New Script in the drop down menu. The path behavior is a script that you can find in the Virtual Sets Resource folder. Drag and drop the behavior on the 3D object called 3D Camera. The path behavior is now applied to the 3D object. You can also create the behav-

Open curve path used inside a 3D object.

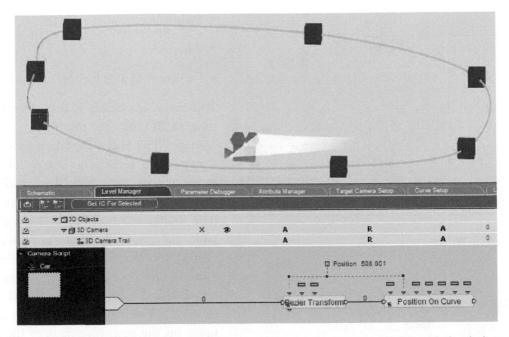

Although the speed of the 3D object can change over time, it is set by a Bezier Transform building block, which controls time, and a Position on Curve building block, which controls the position of the 3D object in function of time. The Position on Curve building block places a 3D object at a specific position on a 3D curve.

ior by selecting the following building blocks: the Bezier Transform building block found under Building Blocks > Logics > Calculator, and the Position on Curve building block found under 3D Transformations > Curve. Drag and drop each building block on the 3D object called 3D Camera.

The script includes Bezier Transformation and Position on Curve.

The Bezier Transform building block is useful to convert a float value into another float value according to the shape of the progression curve. Please note that the float variable coming from the Bezier Transform controls the percentage of Position on Curve. This variable determines the motion of the 3D object on the curve. You can double click on the building blocks to change the following parameters:

- Progression curve

- Name of the curve

- Position on the curve at initial conditions

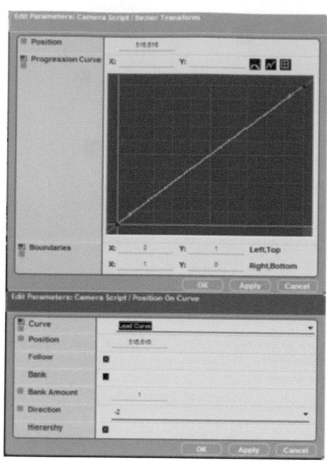

You can double click on the building blocks to change the following parameters: Progression Curve, Name of the Curve, Position on the Curve at Initial Conditions, Follow, Bank, and Direction of Axis for the Forward Motion.

- Follow

- Bank

- Direction of axis for the forward motion

You can assign children and parents attributes to 3D objects inside the Hierarchy Manager. The Hierarchy Manager comes in handy when setting up several objects to move together on a path. The children will then follow the parent object driving on a path. In the example of driving a car on a path, children include the car driver, Ford Cam (the camera located inside the car), the bumper, the doors, and other parts, which are all linked to their parent, the 3D object called Car. The parent object and its children move on the path regardless of the animations of the children. In this example the driver behind the steering wheel can move his or her body and remain a child of the 3D object called Chassis.

2.3   Interactive Motion Control of a Camera on a Path

Goal: The goal of this tutorial is to allow the viewer to interact with an object while in motion on a path. In this example the viewer can control the speed and also some other parameters such as the field of view of a camera. The object can be controlled as it goes forward, goes backward, accelerates, decelerates, and stops. PathSpeed.cmo, the interactive example covered in this section, can be found in the Interactive Paths folder on the companion CD-ROM.

Overview: This tutorial shows how to attach an object on a path and how to control the speed of the camera with a keyboard.

The camera can also decelerate until it reaches a still point when the viewer presses the Arrow Down key. The camera will then go backward, accelerate, and decelerate.

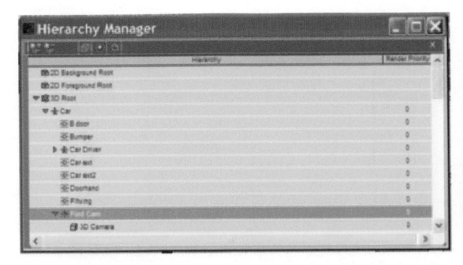

You can assign children and parents attributes to 3D objects inside the Hierarchy Manager. The Hierarchy Manager comes in handy when setting up several objects to move together on a path.

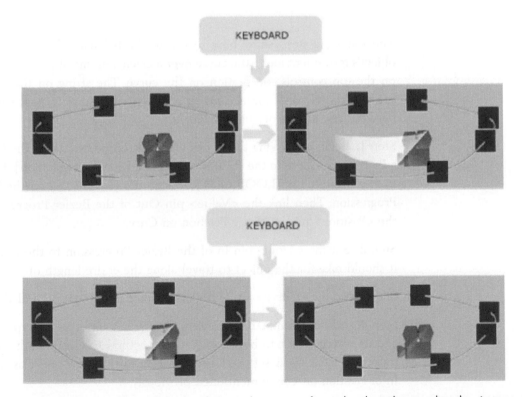

The camera can travel on a circular path, a closed curve where it goes forward and accelerates when the viewer presses the Arrow Up key.

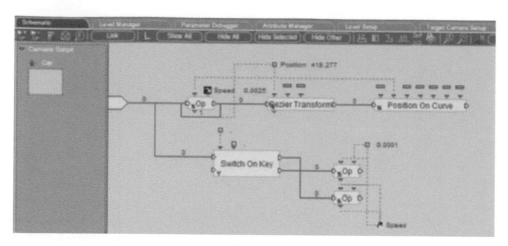

The string on the top controls the motion on the curve. The string on the bottom receives the viewer's input. This tutorial describes how to pass variables between two strings of building blocks.

The following steps show how to create the behavior that controls the object's movement along the curve over a given amount of time. The string on the top controls the motion on the curve. The string on the bottom receives the viewer's input. This tutorial describes how to pass variables between two strings of building blocks.

Step 1. Begin by moving the object along the curve by linking the (IN) of a Position on Curve to the (LOOP OUT) of a [Bezier Progression]. Be sure to also connect the (LOOP OUT) back to the (LOOP IN) on the Bezier Progression. Then link the <Value> pin Out of the Bezier Progression to the <Position> pin In on the Position on Curve.

Step 2. Set the <Time> pin In of the Bezier Progression to the time that it should take for the object to travel along the entire length of the curve.

Step 3. Set the <Curve> pin In of Position on Curve to the intended curve.

Step 4. Add a Parameter Operation. Set its first input to be type Curve and its second input to be None. Set the method to Get Length and the output to Float. Now set the Curve variable to the same curve as the pin In of Position on Curve mentioned previously.

Step 5. Add another Parameter Operation. Set the input and output to Float and the method to Division. Link the output to the Get Length Parameter Operation to the first input of the Division Parameter Operation. Set the second input to the desired speed.

Step 6. Now link the output of the Division Parameter Operation to the <Time> pin In of the Bezier Progression mentioned previously.

Now that we control the motion of a single path, we explore the possibility of jumping from one path to the other.

2.4   Cameras Can Jump from One Path to Another

Goal: The goal of this tutorial is to learn how to script a single camera using several paths and to give the possibility to the viewer of jumping from one path to the other. PathSwitch.cmo, the interactive example covered in this section, can be found in the Interactive Paths folder on the companion CD-ROM.

Overview: This tutorial shows how to switch the script being performed by an object. The Switch Script behavior can activate and deactivate scripts at the viewer's command. This powerful behavior used for interactive

The same path camera jumps from the bottom path to the top path when the viewer presses the space bar.

storytelling is similar to a stage manager giving clues to actors coming on and off of the stage.

In this example, the same path camera jumps from the bottom path to the top path when the viewer presses the space bar. The switch script behavior can activate the top path behavior and deactivate the bottom script behavior. At the next space bar input the sequencer activates the second string of scripts, which deactivates the top path behavior and activates the bottom path behavior. This action brings the camera back to the lower path.

To construct more than two outputs for the sequencer, right click on the Sequencer building block and choose Construct in the pull-down menu. A new output is added on the right side of the Sequencer building block.

2.5   Viewers Can Jump from One Camera to the Other

Goal: The goal of this tutorial is to learn how to script several path cameras and to use a transition camera to move from one path camera to the other camera followed by the viewer. PathTransition.cmo, the interactive example

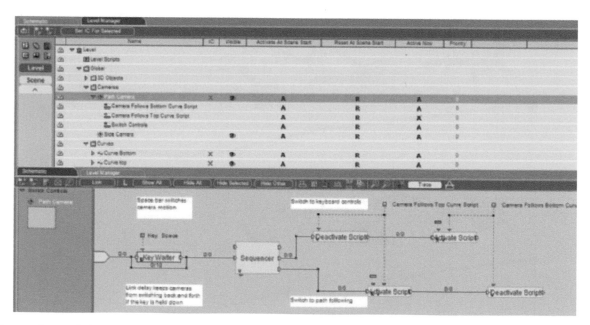

The switch script behavior can activate the top path behavior and deactivate the bottom script behavior. At the next space bar input the sequencer activates the second string of scripts, which deactivates the top path behavior and activates the bottom path behavior. This action brings the camera back to the lower path.

covered in this section, can be found in the Interactive Paths folder on the companion CD-ROM.

Overview: This tutorial shows how to transition between several cameras or 3D objects moving on their respective paths.

The switch between cameras or 3D objects can be a clear cut similar to raw editing of shot cuts in movies or video. You can also maintain the continuity of the narration by using a third camera called Transition Camera, which provides a softer way to create a transition between cameras. The keyboard-controlled Transition Camera will create a path on the fly, start-

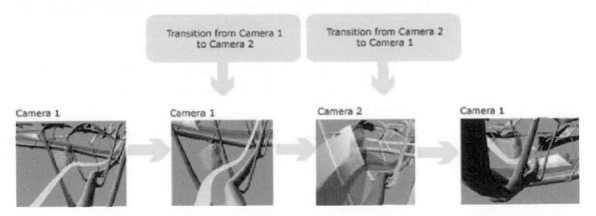

You can maintain the continuity of the narration by using a third camera called Transition Camera, which provides a softer way to create a transition between cameras.

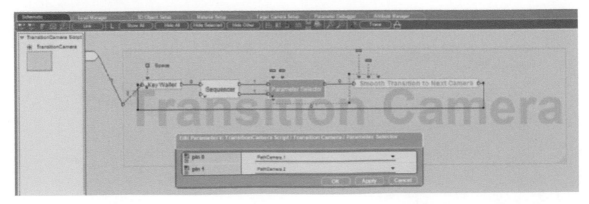

The keyboard-controlled Transition Camera will create a path on the fly, starting at the position of the yellow camera on the yellow path and ending at the position of the green camera on the green path.

ing at the position of the yellow camera on the yellow path and ending at the position of the green camera on the green path.

2.6   How to Switch Cameras on the Same Path

Goal: Viewers can transform a path camera into a free camera. The path camera can be changed **into** a free camera and back to a path camera. This setup is very useful to transform the course of the narration by changing the nature of the camera. Viewers can interrupt the exploration of a path to look around for details. The transformation only changes the behavior applied to the camera by activating a new script. SwitchCam.cmo, the interactive example covered in this section can be found in the Interactive Paths folder on the companion CD-ROM.

Overview: This tutorial shows how to activate and deactivate scripts to transform the behavior of a follow path camera or any 3D object into the behavior of a free keyboard-controlled camera or any 3D object while remaining on a path.

The Key Waiter building block reacts to an input received by the system— for example, a keyboard space bar input—and triggers the sequencer, which deactivates the current behavior for the camera, which is the path camera

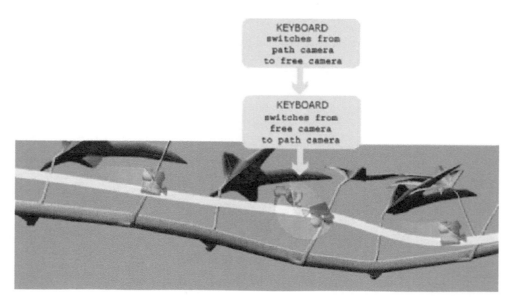

This illustration shows how to use keyboard controls to activate and deactivate scripts to transform the behavior of a path camera into the behavior of a free camera while remaining on a path.

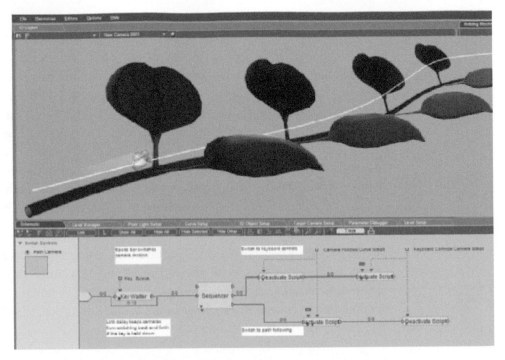

While the path camera follows the light green curve, the Key Waiter behavior is kept on a loop during the ride.

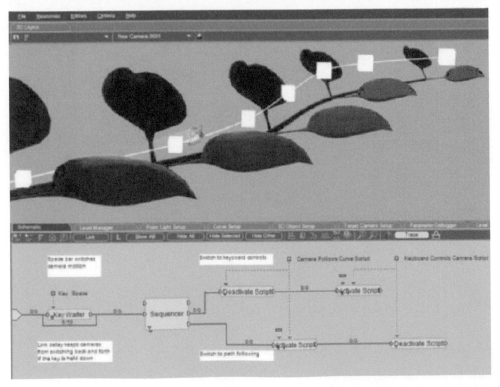

The same camera can go through a path or a free camera sequence depending on the Sequencer building block. In this illustration, the red string deactivates the path script called Camera Follow Path Script and activates the free Keyboard Controls Camera Script.

behavior, and activates a new behavior (in this example the Free Camera behavior). The sequencer works as a switch node that activates the pin next to the pin currently activated. The sequencer activates all the pins down the list until it reaches the last one, in which case it automatically goes back to the first one. In this example you can notice the chains of actions attached to the right of the sequencer, which alternates between the first chain of action, to activate the free camera and deactivate the path camera, and the second chain of action, to activate the path camera and deactivate the free camera.

The Free Camera behavior can freely rotate the camera around its axis. The Switch on Key building block allows the viewer to control the rotation of the camera with the keyboard arrow keys. Several Parameter Selectors building blocks control the angle of rotation along the rotation axis.

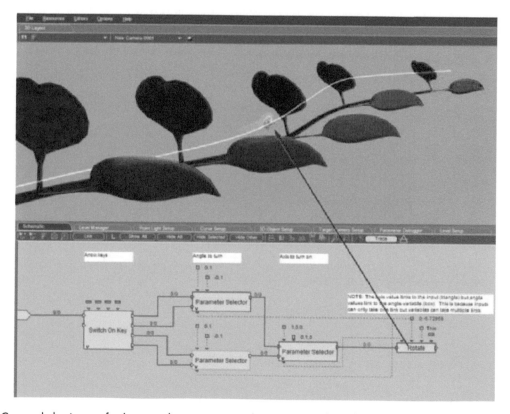

The Free Camera behavior can freely rotate the camera around its axis. Note the red rotation trail. The Switch on Key building block allows the viewer to control the rotation of the camera with the keyboard arrow keys.

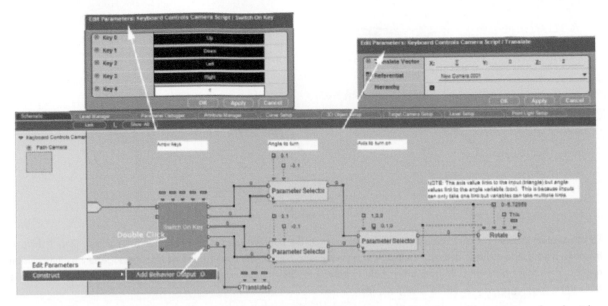

You can construct more outputs around the Switch on Key building block. For example, you can add translation when the 1 key is pressed.

You can construct more outputs around the Switch on Key building block. For example, you can add translation when the 1 key is pressed.

2.7 How a Moving Object on a Path Can Detect and Avoid Obstacles

Goal: The goal of this tutorial is to learn how to script a reactive camera that can switch paths when detecting objects or colliding with objects. For example, a moving object on a path can detect fixed obstacles and avoid them by changing paths. PathCollision.cmo, the interactive example covered in this section, can be found in the Interactive Paths folder on the companion CD-ROM.

Overview: This tutorial shows how to create a collision detection sensor between a camera and objects placed on its path. The camera answers the message sent by the collision event and switches automatically from green path to blue path.

When you hit the Play button, the camera is on green path. After the collision, the camera switches to blue path.

The cube with a Collision Detection building block and the 3D camera exchanged a message called ChangePath. The 3D camera changes from green path to blue path on reception of the message. Please note that using

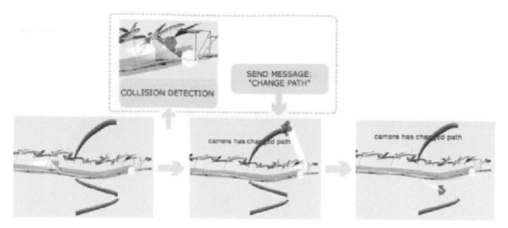

This illustration shows how a collision detection sensor controls the path followed by the camera. The camera, Step 1, follows the path, Step 2. The collision detection sensor detects the camera, Step 3. The camera answers the message sent by the collision event and switches automatically from green path to blue path, Step 4. The camera remains focused on its target, the white ball, while finishing its course on the blue path, Step 5.

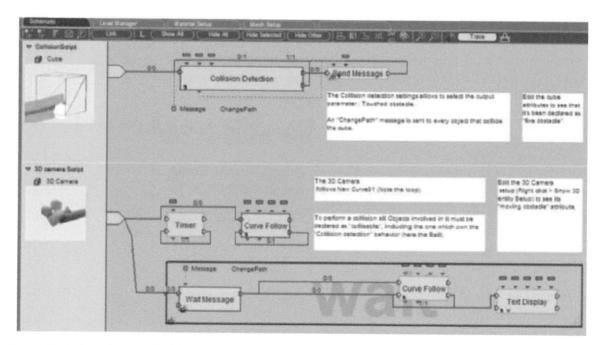

On top, the Collision Detection building block sends a message to a camera or 3D object to signal that it is time to switch paths. The bottom strings have the Wait Message building block. The behavior waits for the message "ChangePath," which will give the signal to change from green path to blue path.

the Follow Path behavior instead of the Position on Curve behavior can help to position the camera at the starting point of the curve when you hit Play. This behavior calls the camera in the starting position, regardless of the initial position of the camera in the world.

In the Schematic view, the difference between both strings is the Wait Message building block on the bottom. The behavior waits for the message ChangePath, which will give the signal to change from green path to blue path. Sending a message to a camera or 3D object signals that it is time to switch paths. Several objects, a ball, a 3D model of a camera, and a virtual camera are attached together to the same parent, a ball moving on a path. The 3D object follows a green path until it is told to switch to a blue path. By changing the Curve Follow path followed by the ball, we can actually switch from the green path to the blue path. In this case both paths are perpendicular to each other. The green path follows the X axis, and the blue path follows the Y axis. In the Schematic view, the top string shows the behavior keeping the camera on the green path. The bottom string shows the behavior keeping the camera on the blue path.

2.8   How to Record a Path on the Fly

This tutorial shows how to generate a curve on the fly that can be saved and reused as a camera path. The curve is recorded as the camera moves through the world. The curve is generated by interpolation between key-frames. For this tutorial, you need to use the file RecordableCamera.cmo available in the CD-ROM. This file will help you to create and save the path that you can reuse as a 3D curve in your project.

The following steps will help you to record a path on the fly in Virtools:

Step 1. Open RecordableCamera2.cmo, and drag and drop your scene's 3D content from your own Data Resource folder under 3D Entities or merge with your own composition (for example, Branch-scene.cmo).

Step 2. Make sure that the camera is in place to start recording the path, and select Set Initial Condition for the camera. You can read the instructions as they come up on top of the screen.

Step 3. Hit Play to activate the recording mode. When you are ready to move the camera, hit the space bar to create the first key-frame.

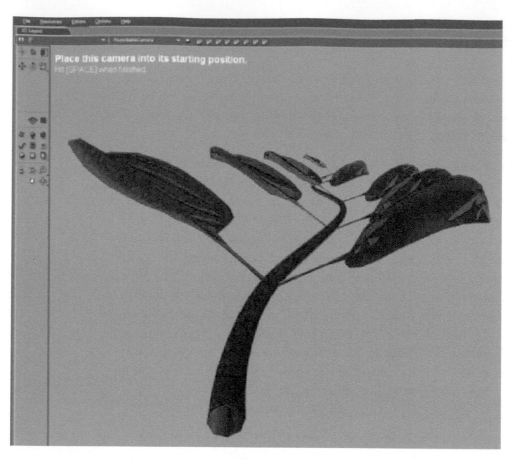

Open RecordableCamera2.cmo, and drag and drop your scene's 3D content from your own Data Resource folder under 3D Entities or merge with your own composition (for example, branch-scene.cmo).

The number of recorded key-frames is indicated on the left of the screen.

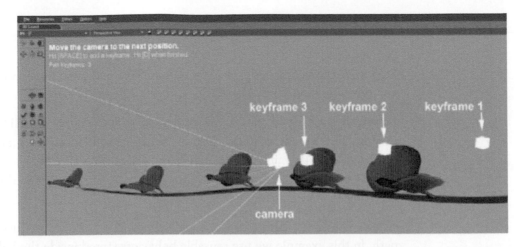

This side view shows the camera being moved to the right. The white cubes are the locations of recorded key-frames.

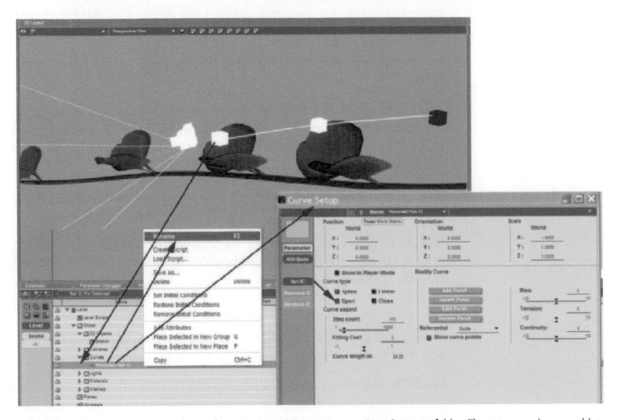

Right click on the curve's name, and save the curve as a 3D entity in your Data Resource folder. The curve can be reused later as a path for the camera by dragging and dropping the curve object in the scene.

Step 4. After recording the last key-frame of the path, you can confirm the curve by selecting Level Manager > Curve > Recorded Curve 1.

Step 5. Right click on the curve's name, and save the curve as a 3D entity in your Data Resources folder. The curve can be reused later as a path for the camera by dragging and dropping the curve object in the scene. The curve will take place in the same location as the curve generated during recording.

After creating the curve, you can add behaviors to 3D objects using the path. In this example we use reusable behaviors stored inside the behavior graph folder in the Virtual Sets data resource. The behavior called Follow Path & Speed, a path camera behavior with interactive speed control created earlier in this chapter, is dragged and dropped on the future path camera.

Setting up a speed control behavior for the path camera takes several steps:

Step 1. In Level Manager, set Initial Conditions for the camera. In Schematic, make the camera idle, turn the speed to zero, adjust the speed of Per Second to zero and turn Position on Curve to zero. Set Initial Conditions for the behavior.

Step 2. Press Play. The path camera gets positioned at the origin of the path and remains idle.

Step 3. Double click on Position on Curve, and adjust the axis for the camera. Hit Play, and the path camera glides on the path. Hit the keyboard arrow keys to control speed.

3   THE DONUT TUTORIAL: SETTING UP A CAMERA INSIDE AN INTERACTIVE RACE TRACK

Goal: The following tutorial is a step-by-step creation of an interactive race track inside a virtual donut. The tutorial shows how to model a race track inside the interior space of the donut and how to set up cameras racing on closed curve paths. The tutorial presents interactive functions for cameras on a path including changing direction, controlling speed, and field of view. Some sections of this tutorial can be bypassed in case you want to reach a specific level of functionality.

View of a camera racing inside a textured donut. The red trail left behind the camera is a visual indication of the speed of the camera.

The Lightwave and Virtools files for this tutorial can be found on the companion CD-ROM. A step-by-step tutorial on building a donut in Maya and setting up the path inside the donut in Virtools can be found on the CD-ROM under Interactive Paths > Tutorials > donut-maya.pdf.

3.1   Conceptual Design

This tutorial shows how to build the donut from scratch. The following summary will help you visualize the steps described in detail in the tutorial.

1 Modeling a donut-shaped ring. Although we use Lightwave for this tutorial, donut shapes are available as 3D primitives with most 3D modeling applications. The tutorial also shows how to twist a donut to create a complex Moebius ribbon. Keep in mind that from a viewer's point of view, the environment and the obstacles encountered during the race may be more important than the shape of the race track.

2 Texturing the donut. Textures placed inside the donut are part of the interactive experience. Attractive textures help to keep the viewer interested in the content—for example, kinetic textures provide visual effects that can change when seen at various speeds. These textures will provide

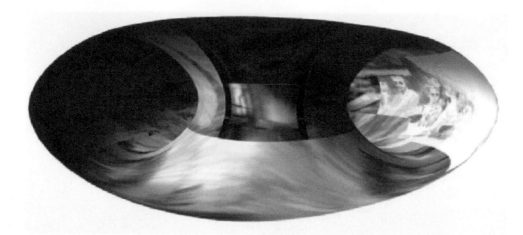

Textures placed inside the donut are part of the interactive experience.

different sensations to the viewer when the camera accelerates and decelerates.

3 Adding lights and cameras. After setting up a scene with lights and cameras in Lightwave Layout, the tutorial shows how to add interactive functions and behaviors in Virtools. The following steps cover simple interactive behaviors—for example, how to set up a follow camera on a ring and how to create an environment that can react to the camera.

4 Creating interactive textures.

5 Creating an interactive path.

6 Having one camera follow the path.

7 Controlling the camera's speed on the path.

8 Switching between the path camera and the free camera.

9 Adding an additional view to follow the racing cars on the circuit.

10 Designing an interface with buttons to toggle between several cameras.

11 Controlling interactive textures according to the camera's speed.

12 Switching between paths.

13 Mounting cameras inside cars.

14 Creating car collisions with unpredictable moving objects.

15 Editing interactive paths.

3.2   Let's Start the Tutorial

3.2.1   Conceptual Design

The phase of conceptual design helps to sketch simple ideas regardless of the technology or the process used to implement your ideas. The drawings

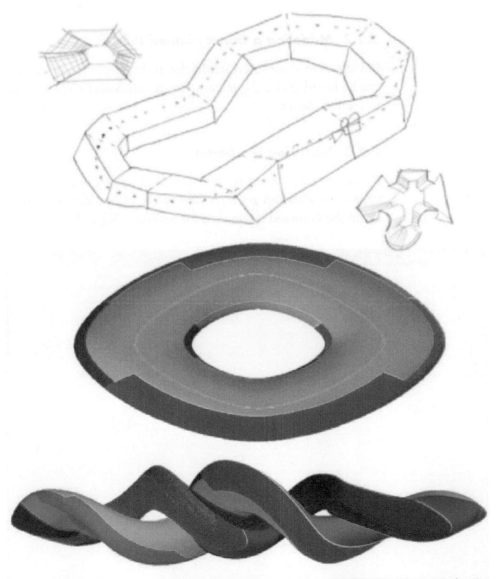

Top view: Sketches exploring ideas for the ring and various sections of donuts. Bottom view: 3D model of a donut and of a triple Moebius ring modeled from a donut.

on page 245 explore several segments of the rings and ways to model the circuit with various sections of tubes.

The tutorial will show you how to reuse a setup created for the simple donut for a more complex ring such as a triple Moebius ring. It is always easier to test behaviors on a donut and to later duplicate the same behavior inside a more complex model.

3.2.2 Modeling a Donut-Shaped Ring

In Lightwave Modeler, create the donut shape, which is a 3D primitive that can be found under the menu's Create tab. Select Create > Superquadric to create a donut.

3.2.3 Texturing the Donut

By selecting the donut in the Nurb mode, you can make the skin of the volume smooth and get a preview of the object similar to what you will get in the Gouraud mode used for real-time 3D in Virtools.

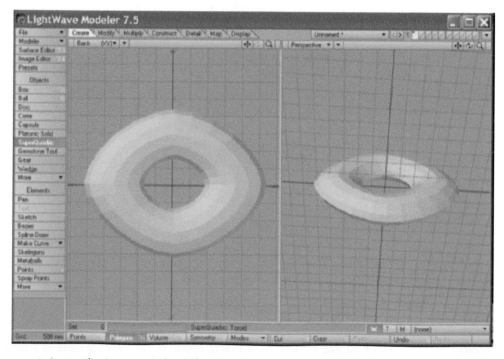

Although we use Lightwave for this tutorial, donut shapes are available as 3D primitives with most 3D modeling applications.

Select surfaces with names and colors that will receive textures for the race track. Please note that Virtools used for this tutorial assigns names to materials according to surface names created in Lightwave and materials in 3Dmax. The look and feel of the circuit will depend on the type of projection—cubic, planar, spherical, or cylindrical—chosen for the textures.

Please note that textures may be replaced in Virtools by another texture, by a video clip, or can be combined with another texture. Because we will add interactivity to existing textures to scroll images, to map video textures, or to blend with other textures, interactive textures created in Virtools will use the same texture coordinates as the 3D software where they have been created. The original texture assigned to a surface in Lightwave, Surface Editor, will be used as a default texture when imported in Virtools. The default texture will be altered or modified in Virtools.

Various names and colors are assigned to surfaces all around the donut. In Lightwave Modeler, select the surface by going to Details > Polygons,

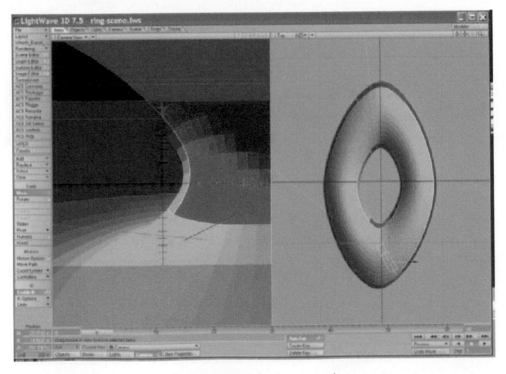

Select surfaces with names and colors that will receive textures for the race track.

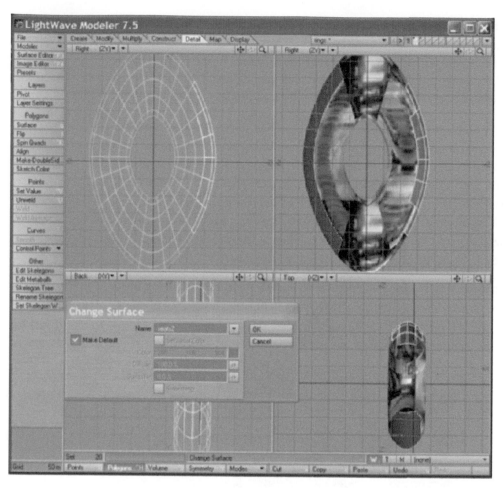

The selection process covers the inside and outside of the ring. After texturing, create a culling effect by flipping all the polygon normals.

Surface. The selection process covers the inside and outside of the ring. After texturing, create a culling effect by flipping all the polygon normals.

The inside of the donut is now fully textured and the donut is transparent when seen from the outside.

3.2.4 Adding Lights and Cameras

In Lightwave Layout, we add cameras and lights placed at interesting locations on the set.

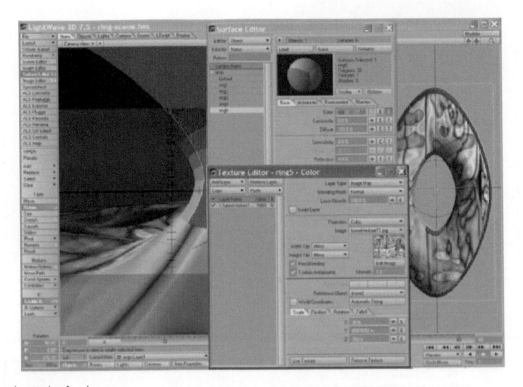

Texturing the inside of a donut.

In Lightwave Layout, creating several cameras located inside the virtual set helps you to choose hot spots that will be available during the production process of your virtual set. Some of the cameras created inside the donut will later receive behaviors and will be active in Virtools.

The file is saved as an .nmo file, which is Virtools' 3D object format. Please make sure that the textures used for this project are also included in the Textures subfolder.

3.2.5 Drawing a Path

Let's create a path in Virtools. Select 3D Layout > Top View. Select the curve tool on the left of the screen. Click in the 3D Layout window to position four new points on each side of the donut with Closed Spline checked. Please note that you can edit a path by moving existing points or by inserting additional control points. Paths can be duplicated if you need to guide other 3D objects—for example, lights, cameras, and their targets.

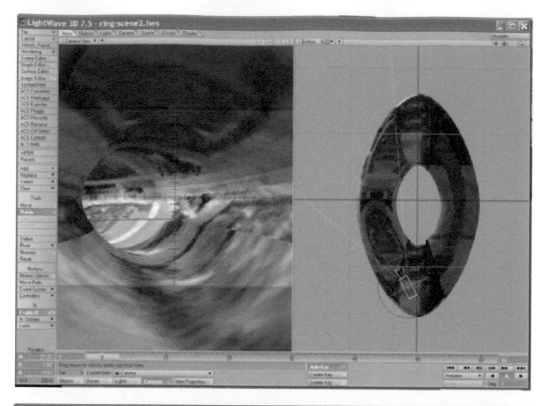

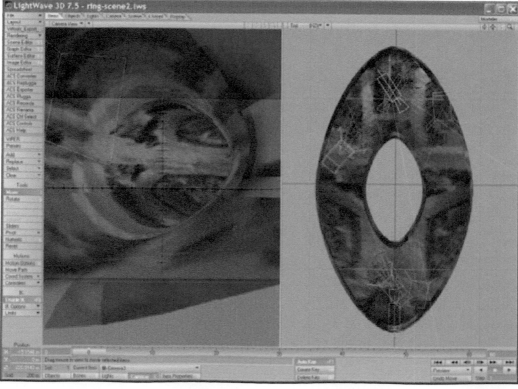

Top view, left: The point of view from camera 1. Right: A top view of the donut showing camera 1 with rotation handles selected. Bottom view: Point of view of camera 2 after adding red spotlights.

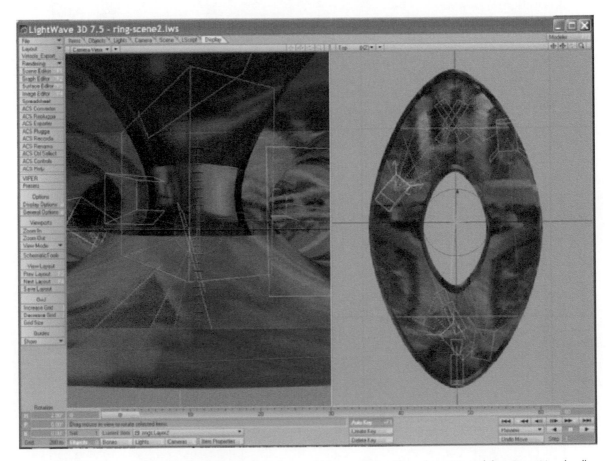

Point of view of camera 3 after adding yellow, purple, green, and blue spotlights. Please keep in mind that OpenGL only allows rendering eight lights in a single camera shot.

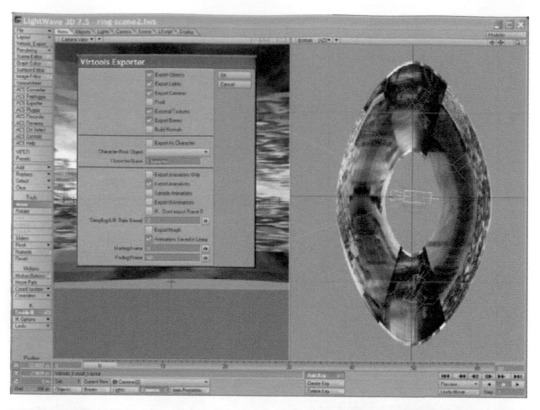

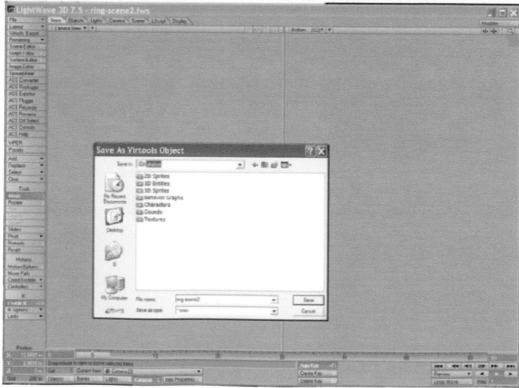

Top view: In Lightwave Layout, a textured model with lights is ready to be exported to Virtools. Bottom view: The model of the donut is saved inside the 3D Entities subfolder of the Virtools resource folder.

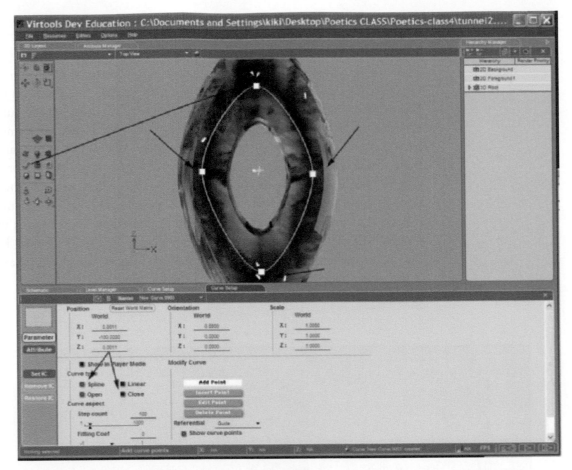

Let's create a path in Virtools. Select 3D Layout > Top View. Select the curve tool on the left of the screen. Click in the 3D Layout window to position four new points on each side of the donut with Closed Spline checked.

3.2.6 *Creating an Interactive Path*

Please note that the Follow Path behavior can be found in the Virtual Sets, Resource Folder > Behavior Scripts, located on the right of the screen. You can access this behavior, and many other behaviors created in this book, from the companion CD-ROM. Open Virtools, go to the top menu, select Resources > Open Data Resource, and select the file virtual sets.rsc on the CD-ROM. On the right side of the Virtools interface, select Behavior Graphs > Interactive Paths > Follow Path. This behavior was created earlier in section 2.1 of this chapter. You can refer to this section for more details about how to design this behavior. Although we decided to apply the behavior to a camera in this example, you can follow the same steps with other 3D objects or interactive characters. Go to Level Manager > Global > Camera, and drag and drop the Follow Path behavior on the word "camera."

The script is added under the camera named Path Camera in this example. Double click on the script's name to open the Schematic window with the Follow Path script. Edit the Position on Curve with the name of the curve used as a path. When testing the scene you will notice that the camera follows the continuous curve without interruption. You may want to change the position and orientation of the camera on the curve. Stop the playback, and go to Initial Conditions. Select and move the camera to a new position. Select Initial Conditions for the camera, and hit Play. The camera follows the curve with a new position and orientation. You can edit the direction of the camera by selecting the reference axis of the object following the curve. For example, if the camera moves forward along a Z axis, it will move backward when you select the −Z axis.

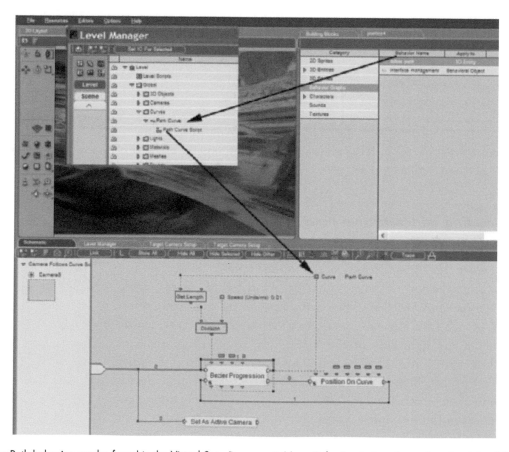

The Follow Path behavior can be found in the Virtual Sets, Resource Folder > Behavior Scripts, located on the right of the screen.

3.2.7 Controlling the Camera's Speed on the Path

The viewer can control the speed of the camera. Please note that the Follow Path & Change Speed behavior can be found in the Virtual Sets, Resource Folder > Behavior Scripts, located on the right of the screen. Please note that this behavior was created earlier in section 2.3 of this chapter. You can refer to this section for more details about how to design this behavior. Open Virtools, go to the top menu, select Resources > Open Data Resource, and select the file virtual sets.rsc on the CD-ROM. On the right side of the Virtools interface, select Behavior Graphs > Interactive Paths > Change Speed. Go to Level Manager > Global > Camera, and drag and drop the Follow Path & Change Speed behavior on the word "camera." The script is added under the camera named in this example Path Camera. Double click on the script's name to open the Schematic window with the Follow Path script. Edit the Position on Curve with the name of the curve used as a path. After setting up the Follow Path & Change Speed behavior scene, you will notice that you can accelerate and decelerate the camera by pressing the arrow keys on your keyboard.

3.2.8 Switching between Path Camera and Free Camera

The viewer can use the keyboard to turn a path camera into a free-range camera. We can add a new interactive behavior to the camera by switching between the camera scripts. Please note that the Free Camera script and the Switching Script script can be found in the Virtual Sets, Resource Folder > Behavior Scripts, located on the right of the screen. Please note that these behaviors were created earlier in section 2.4 of this chapter. You can refer to this section for more details about how to design this behavior. Go to Level Manager > Global > Camera, and drag and drop the Free Camera script and the Switching Script script on the word "camera." The scripts are added under the camera named Path Camera. Double click on the script's name to open the Schematic window with the Switching Script script. In the top line, edit the names of the Follow Path Camera and the Free Camera inside the Activate and Deactivate building blocks. After setting up the Switching Script script and the Free Camera script, you will notice that you can press the space bar on your keyboard to turn your path camera into a free camera and vice versa.

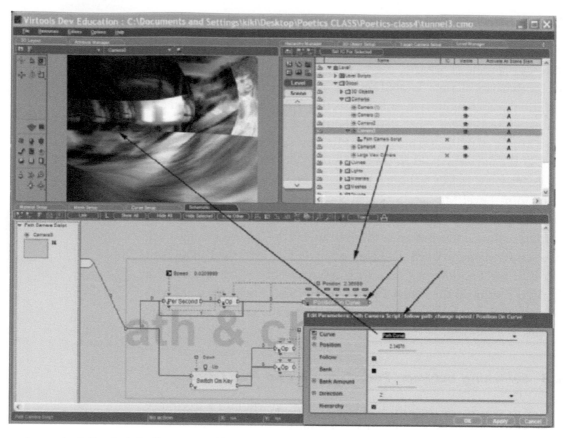

After setting up the Follow Path & Change Speed behavior scene, you will notice that you can accelerate and decelerate the camera by pressing the arrow keys on your keyboard.

3.2.9 Creating Interactive Textures

The viewer can control and/or replace textures applied to the inner and outer rings with scrolling images or videos or by adding interactive blending in relationship with variations of the camera's speed. In this example, we use procedural textures with interactive behaviors described in detail in the chapter about interactive textures (Chapter 3).

Please note that the Background is a procedural texture created in the chapter about interactive textures. In Virtools, the Background texture and its behavior can be found in the Virtual Sets, Resource Folder > Textures Folder, located on the right of the screen. Before setting up the procedural texture, you need to drag and drop the Background texture in the 3D Layout window.

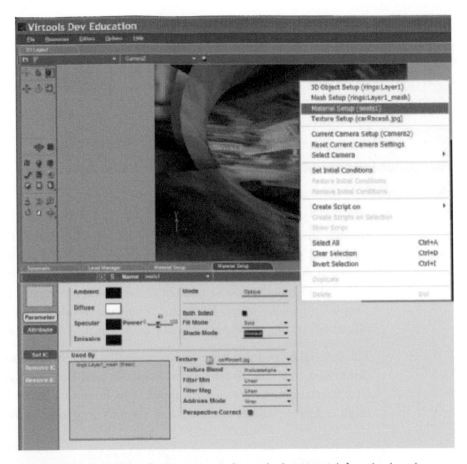

In Virtools, right click on the donut viewed in the 3D Layout window and select Materials from the drop down menu. The Materials Setup window opens in the lower part of the screen.

In Virtools, right click on the donut viewed in the 3D Layout window and select Materials from the drop down menu. The Materials Setup window opens in the lower part of the screen. Inside the Materials Setup window, select the Texture pull-down menu and replace the existing texture with a new texture called Background.

3.2.10 Adding an Additional Top View Helps to Keep Track of the Locations of Cameras on the Circuit

The viewer can follow the race with an additional top view created inside the screen. In Virtools, go to 3D Layout and select Top View. Create a new

The viewer can follow the race with an additional top view created inside the screen.

camera. In Level Manager, rename the camera Top View. The behavior Additional View can be found under Building Blocks > Interface > Screen. Drag and drop the behavior on the words "Top Camera" in Level Manager > Global > Camera. Double click on the script created under Top Camera. This will open the script inside the Schematic window. You can edit the placement and size of the additional top view inside the 3D Layout screen.

This tutorial helped you to design an interactive experience made of several parameters, textures, depths of field, speeds, and camera angles that can change according to the viewer's input. This simple scene can be reused in more complex projects—for example, a car driving inside a virtual city.

CHAPTER 6
Virtual Cameras

1   INTRODUCTION

The design of a virtual world and the way that you film a virtual world are different ways to influence a viewer's experience. The use and design of interactive cameras are key elements of telling stories in virtual spaces.

Cinema brought us the possibility of seeing and experiencing a scene through someone else's eyes. Moviegoers are invited to view the world from the viewpoint of a film director or director of photography, but they do not have the ability to change the director's viewpoint. They cannot choose, for example, which of multiple points of view they want to use for a scene.

In virtual sets, viewers also become authors because they can actually direct and create additional viewpoints for a scene using virtual cameras. Viewers in virtual worlds can also take advantage of interacting with the scenes they are viewing by changing the order of time and space inside a scene. These new possibilities of virtual cameras and interactivity allow viewers to revisit some of the conventions of cinema. For example, the traditional division of tasks between shooting and editing becomes obsolete with virtual cameras that can process both activities at the same time.

Virtual cameras are designed to film virtual worlds without the need for additional editing. The viewer is the author, director, and editor. Virtual cameras are located everywhere and they can film all the time. Viewers can control them, or the virtual cameras can make their own decisions. Independent virtual cameras can be programmed to make decisions and adapt themselves to different types of content. They can follow the topography of the scene being filmed and modulate how they shoot and frame unpredictable events.

This chapter covers the steps necessary to create virtual cameras that are fully interactive and autonomous. The conceptual design and practical implementation of virtual cameras are explained step by step, starting from basic cameras that exhibit simple behaviors and expanding to self-determined cameras with increasingly complex behaviors. The chapter starts with virtual cameras that share the same attributes and behaviors that movie cameras possess. By looking at the role of cameras in scenes from classic movies created by film directors such as Hitchcock, Antonioni, and Kubrick, directors whose styles can be easily identified, we can design virtual cameras that will emulate these familiar camera styles and resulting effects for viewers of a virtual world.

1.1   Relationships between Virtual Cameras and the Environment of a Scene

This chapter will explore relationships between virtual cameras and the environment of a scene. We will show how creating an environment for a scene can be a crucial element that affects the scripting of virtual cameras. We will try to catalog cameras in several groups sharing the same behaviors. This will help us to outline groups of virtual cameras built around similar behaviors.

Cameras covered in this chapter include the following:

- Follow camera

- Over-the-shoulder camera

- First-person camera

- Third-person camera

- Dialog camera

- Transitional camera

- Dolly camera

- Orbital camera

- Chopper camera

- Jet camera

- Car camera

Filming inside a virtual world with virtual cameras requires technical skills and some psychological insights. The technical properties of virtual cameras

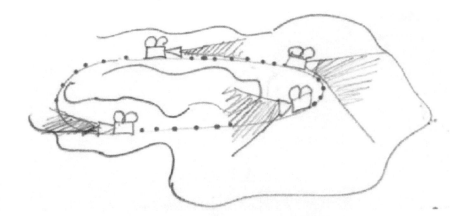

A virtual camera following the looped path of a race car can change the depth of field with speed. The cone of vision of the camera is controlled by the speed of the camera. As a result, the perception of the landscape will change.

can be expressed in terms of attributes and scripts like any other 3D objects in a virtual world. Virtual cameras also have specific parameters and behaviors that help them to react and change the psychology of a scene. For example, a virtual camera following the looped path of a race car can change depth of field with speed. This effect of speed will create a deformation of the landscape for a viewer. The interactive demo for this virtual camera can be found on the companion CD-ROM in the Interactive Camera folder under PathFOV.cmo.

In this example, the virtual camera remains the same but the camera's attributes can change the perception of the environment. The scenario of this little race car example covers someone driving a car at high speed around a race track, stopping the car at a certain point, getting out of the car, and then walking around the vehicle. The following details continue to address the elements used in this virtual scene.

Imagine being able to switch from a dolly camera mounted on tracks to a steady camera mounted on a wheelchair without the need to change the equipment. In this case a virtual camera can change behavior and become another kind of camera by switching scripts according to the context of a scene.

A path or chase camera can switch behaviors when the car acts a certain way or stops. For instance, a sensor can detect a camera entering inside the blue area shown in the illustration on page 263. The sensor sends a signal to switch the camera's script. The path camera then takes on the role of a free-range camera. The viewer also experiences a change of camera when the driver gets out of the car.

This dolly camera mounted on tracks is the equivalent of a virtual camera on a path.

This camera mounted on the seat of a wheelchair is the equivalent of a free-range virtual camera.

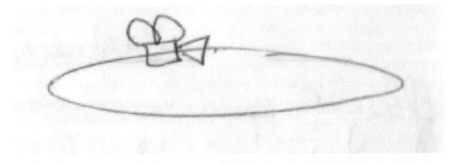

Camera on a looped path.

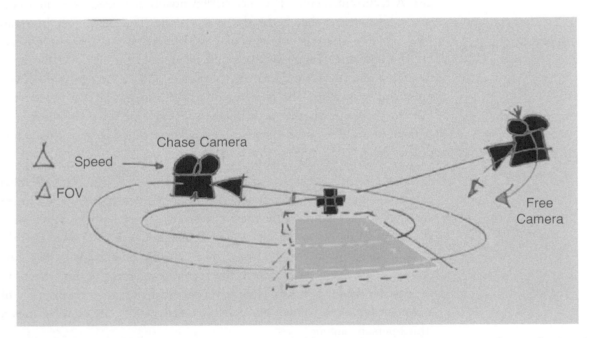

The path camera can switch behaviors when the car stops in the blue area. The camera triggers a sensor that switches its script when the camera enters inside the blue area. The path camera turns into a free-range camera. The viewer experiences a change of camera when the driver gets out of the car.

1.2 The Psychology of the Viewer

The design of virtual cameras that can make their own moves and decide how to frame characters in a scene requires the introduction of behaviors that can ensure the stability of the image with slow transitions while ensuring a sense of continuity to keep viewer's interest. Several psychological issues discussed in this chapter are related to the need to stabilize the image inside a frame and to create continuity between frames seen by the viewer.

Another psychological aspect associated with cameras in a virtual space is the way characters in the virtual world can actually acknowledge the camera's presence. The camera can be more than a simple observer of a scene. Virtual cameras can provide the necessary information so the viewer can actually jump in and interact with the scene. Anthropologists, sociologists, and linguists are becoming more aware of how their own presence changes the very nature of what they observe. This can be summarized with the story of an anthropologist who met people handing out questionnaires that they had already filled out during the visit of a previous anthropologist. A recurrent theme of science fiction novels is a newcomer on a new planet discovering that he or she is not alone. The newcomer becomes aware that he or she is observing an environment that in return is observing the observer. This subjective point of view of the observer has been described and filmed in documentary movies and in footage shot by war press correspondents. The observer—or, in this case, the eyes behind the camera—becomes another actor in the scene. He or she is an active participant in the heart of the action and is acknowledged by the other actors of the scene.

Man with the Movie Camera from director Djiga Vertov is a 1928 Russian fictional documentary about a day in the life of a cameraman. The camera and the eyes behind it are the main characters of this masterpiece showing all aspects of the life of his city. One famous sequence from *Man with the Movie Camera* is an early example of a car chase, filmed with two cars driving side by side. The team in the second car films the cameraman in the first car who is filming people in the street. This is an early example of a camera filming a camera and what we call in 3D interactive games a "third-person camera."

One famous sequence of *Man with the Movie Camera* from director Djiga Vertov, filmed with two cars driving side by side, is an early example of car chases. The team in the second car films the cameraman in the first car who is filming people in the street.

This example shows the intimate relationships that form between the observer, the eye behind the camera, the action, and the environment of the scene. In early single-person shooting games in which the camera is associated with a gun, shooting means seeing. This is an extreme example of a first-person camera.

In more traditional storytelling, the story teller/observer/voyeur is unseen by the actors of a scene and remains passive or invisible as the story is being told. This point of view of the observer being removed from the action comes from a long tradition of theater presently portrayed in television sitcoms in which the audience enjoys a front view of the stage being filmed.

In the book *One Thousand Nights*, the narrator tells the reader first-hand accounts but mostly shares stories heard from other narrators. This narrative structure is not entirely a first-hand story but a second- or even a third-hand story that was passed through many people before becoming the story of the narrator who tells it to us. The book of *One Thousand Nights* shapes another kind of relationship between the environment and the story. A story in the story from *One Thousand Nights* can be compared with a virtual camera filming another virtual camera filming the world. In the following screen shot from the installation called "Nighthawks," we discover a story being told inside another story. The interactive demo for this virtual camera can be found on the companion CD-ROM in the Interactive Camera folder under Nighthawks.vmo.

On the left view a character is being filmed, and on the right view we can see the camera filming this actor. These multiple possibilities offered by virtual worlds are very powerful when cameras can choose their characters and become semiautonomous. (For more details please refer to Chapter 7.)

1.3   Cameras Are Rarely Working Alone

Whether filming from a moving truck, from a wheelchair, or while running with a handheld Steadycam, cameras are rarely working alone. Looking back at the history of cameras and the use of cameras by directors of photography, we find many examples of cameras working in conjunction with lighting and other portable elements including light sources, motion stabilizers, and monitors of the action being shot.

Lighting a character riding a horse requires following the camera while shooting in motion. In early times this type of activity was tricky, when

In this screen shot from the installation called "Nighthawks," we discover a story being told inside another story. On the left view a character is being filmed, and on the right view we can see the camera filming this actor.

extra electric power required for large light sources could not be provided sufficiently by generators on a truck. The technical problems that surfaced as a result of cameras and lights moving on trucks were resolved with gaffers carrying electric cables. The length of the power cables tended to dictate the duration of the sequence. Virtual cameras do not face the same limitations because they can be designed with lights attached. In the case of a character moving with his or her own light source and a camera attached to the character, the camera can also follow animated targets moving with

Character moving with his or her own light source. The camera can follow animated targets also moving with the character.

the character. Spotlights, targets, and cameras all become part of the moving character.

In the case of a car driving at night, the headlights may point in a different direction than the line of sight of the driver. A camera looking over the shoulder of the driver will follow a different target than the car headlights. This difference of location between the camera's target and the headlights' target creates an interesting tension that builds another source of content for a story.

The graphic above illustrates the relationships between the virtual camera, the character, and the light. We can use schematic diagrams to understand how the elements relate together. This is very helpful when we need a conceptual sketch for programming a new action or behavior.

Virtual worlds offer a unique opportunity to recreate visual relationships between the observer, the action, and the scene. There are occasions when the virtual camera is associated specifically with a character or another

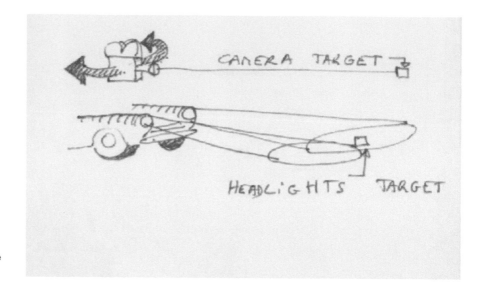

While driving at night, car headlights may point in a direction that differs from the line of sight of the driver.

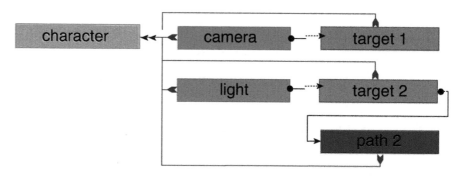

The elements associated with the virtual camera, lights, and character are plotted in a schematic diagram that can be used as a conceptual sketch for programming behaviors.

element of the scene—for example, lighting. In the movie *Raging Bull*, film director Martin Scorcese explores the design of camera angles where the viewer is placed in the situation of one of the boxers and is taking part in a fight. We can actually sense the opposing boxer coming toward us. Another example of a relationship between the virtual camera and a character can be found in scenes from Steven Spielberg's *Indiana Jones and the Temple of Doom* shot in a mine cart riding at full speed inside a mine. In this case, the observer, camera, and character are moved together on a roller coaster that takes the viewers from one scene to another.

In this process the path of the camera is prepared for the viewer who has very little control of where the story goes. The camera is literally the vector of the action—a dynamic storytelling element of a scene attached to the flow of the action.

The trio of the observer, camera, and character is moving on a roller coaster, allowing the viewers to move from one scene to another.

2 BUILDING VIRTUAL CAMERAS

We are now going to look at ways to combine technical and psychological behaviors to design and build virtual cameras that can react to their environments. For example, we will create a virtual camera called a "cocktail camera" that can keep track of certain moving characters and decide to switch from one character to another if the character is leaving the "cocktail party." These cameras can send automated messages exchanged through a virtual world. The messages can reach a group of characters to tell them

A virtual camera called a "cocktail camera" can keep track of certain moving characters and decide to switch to another character if the person being filmed is leaving the "cocktail party."

 to look at the camera. The interactive demo for this virtual camera can be found on the companion CD-ROM in the Interactive Camera folder under CocktailCamera.cmo.

Virtual cameras such as the cocktail camera can be interactive or autonomous and self-controlled with the ability to shoot and edit content without requiring the viewer's input. They are designed to anticipate events and film automatically. Virtual cameras can interact with their environment by tracking certain characters, or they can act like a character moving inside the environment of a scene. Interactive behaviors can be customized to create a self-sufficient camera with a stylized way to look at the world. Several behaviors applied to a virtual camera create more complex attitudes and forms of intelligence that allow the viewer to watch real-time documentaries and fictional movies as they are being shot inside virtual spaces.

2.1   Motion Stabilization

The desire to freely film events while wandering through a space has generated many projects with handheld cameras. Unless the desired effect is that of an experimental movie, the need to record while walking or following someone comes with some obstacles and may create some image-stability problems for the viewer.

The stability of the frame seen by the viewer is a critical element of the success of your content. Behaviors designed to stabilize a virtual camera also set out to accommodate technical and psychological parameters. Stabilization behaviors are also designed to add suspense and other psychological elements to a frame. For example, the delay or anticipation of an action taking place inside a frame can be tied in or associated with camera movements.

In the sketch on page 272, the blue camera is following a character. The blue camera's reaction to the character stopping suddenly is to bump into the character's back. By incorporating stabilization behavior the blue camera can respond to the sudden stop by sliding on the character's back and orbiting around him. The viewer will automatically discover a front view of the character's face, providing visual information about the character's expression. In the case of the red camera, there is no way to anticipate the character's sudden

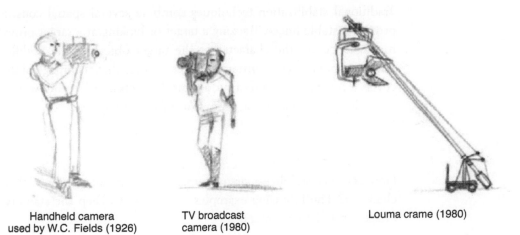

Handheld camera
used by W.C. Fields (1926)

TV broadcast
camera (1980)

Louma crame (1980)

Sketches of major improvements of stabilization for cameras in motion from left to right: Handheld camera used by W.C. Field in 1926; TV broadcast video camera carried on the shoulder around 1980; and Louma crane with remote control camera for panning and tilting.

In this sketch, the blue camera is following a character. The blue camera's reaction to the character stopping suddenly is to bump into the character's back. In the case of the red camera, there is no way to anticipate the character's sudden turn. The red camera shows its surprise by slowing down on its path and almost coming to a stop while looking at the moving character. The red camera will soon follow the character in the other direction but with a delay.

turn. The red camera shows its surprise by slowing down on its path and almost coming to a stop while looking at the moving character. The red camera will soon follow the character in the other direction but with a delay.

Traditional stabilization techniques combine several spatial constraints to provide a stable image. Tracking a target or looking at a target provides the means to center the character or the target object in the middle of the frame. Remaining at a constant distance ensures that the image of the target covers a constant amount of surface inside the frame. In other words, it prevents dramatic zoom in or zoom out effects to happen.

2.2   Virtual Cinematography

How does one stabilize a virtual camera during its interaction with a virtual character? The following examples show how to keep the stability of the image for a virtual camera.

The files illustrating these examples can be found on the companion CD-ROM.

2.2.1 *Character-Dependent Camera*

The first example is a Character-Dependent behavior that keeps the camera attached to a character. The second example shows a camera behavior that is independent from any character but attached to a character from time to time. This behavior also allows the camera to decide to move from one character to the other.

This sketch is an example of this technique. The camera is kept on a leash, like that used for a dog, called Look At. The leash is attached to the character, but the camera can move freely within a certain radius around the character. This method imitates a gyroscopic system that can provide a relatively stable image around a moving character. This camera system moves with the character and is character dependent. We are going to look at ways to create a virtual orbital camera system that can decide when it wants to be dependent or independent from a character. The virtual camera sometimes will be attached to a character; at other times it will be independent from the character, which imitates the situation of someone going from one person to the other during a cocktail party. The version of the stabilization system shows how the Keep at Constant Distance building blocks connected to a proximity sensor can check to see if a new character is walking by. This adaptation of the Looking

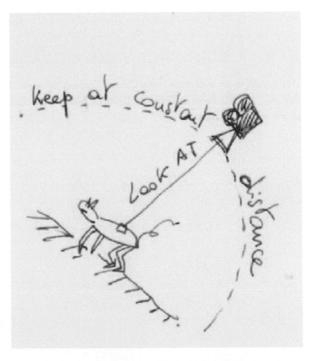

Traditional stabilization techniques combine several spatial constraints to provide a stable image.

This camera system moves with the character and is character dependent. This diagram represents the Orbital Camera behavior, which imitates a gyroscopic system framing a relatively stable image around a moving character.

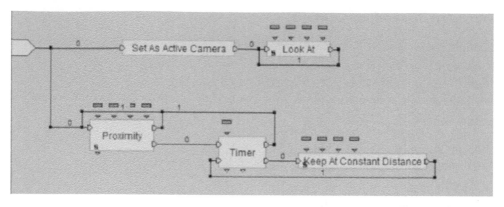

This system is an evolution of the Orbital Camera behavior. In this case the camera is independent from the characters. You can see how the Keep at Constant Distance building blocks are connected to a proximity sensor that checks to see if a new character is walking by.

The Steadycam enables a cameraman to have more flexibility while filming a character. A virtual camera that uses a proximity sensor can establish a similar type of filming flexibility.

At and the Keep at Constant Distance building blocks provides a more dynamic camera system that can adapt to changes of distance and orientation between the character and the camera. The interactive demo for this virtual camera can be found on the companion CD-ROM in the Interactive Camera folder under GyroCamera.cmo.

2.2.2 Camera System Independent from Characters

In this example we are trying to separate the motion of the camera from the motion of the character that the camera is following. A proximity sensor is added to the virtual camera to slow down the reaction of the camera, adding a few seconds lag between the virtual camera and the character. The interactive demo for this virtual camera can be found on the companion CD-ROM in the Interactive Camera folder under FollowCamChasebird.cmo.

This addition of a small delay before the camera starts moving puts the viewer into a mode of anticipation and prepares him or her for the motion that is about to begin.

The virtual camera uses a proximity sensor, making it similar to a Steady-cam. This camera follows a character with a slight delay, allowing the viewer to feel the direction of the character before the camera starts to move. The camera orbits slowly around the character when it stays within a certain distance of the camera. If the character moves past that distance, the camera will start to follow the character. This advanced camera can keep track of two characters moving inside a scene by framing the scene so both characters are included at all times.

At the beginning of the scene, the camera orbits slowly around the characters. You can notice in the following four views that the camera slowly orbits the environment while the characters are standing at a distance in the same space. The interactive demo for this virtual camera can be found on the companion CD-ROM in the Interactive Camera folder under AutonomousCamera.cmo.

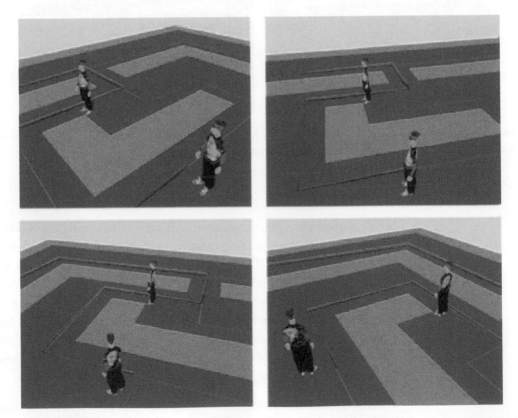

Autonomous camera orbiting around still characters. This sequence of four frames shows the camera's slow orbiting motion while the characters are waiting in the same space.

Autonomous camera with moving characters. This sequence of four frames shows the camera's slow orbiting motion while the characters are moving to reach their goal in the center of the scene.

In this second example, the autonomous camera made decisions about how to film two characters walking toward the same goal without receiving any input from the viewer. This virtual camera is constantly adjusting its position while including the two characters inside the frame. Please note that this scene has no predefined path created for the characters. The characters, like the camera, are fully autonomous. Please check Chapter 4 for more information about self-determined characters. To reach their goal, the gray object in the center, the characters are creating the path where they want to go using path finding, a technique described in Chapter 7.

There are a number of steps that the camera completes to create these four simple frames. The camera's behavior is designed around a sequencer that can activate several stabilization behaviors in a specific order. First, the camera runs tests about what is seen inside the frame. Afterward the camera adjusts when some characters are not included in the frame. It does this by using View Frustum, a building block from Virtools. View Frustum is a behavior that will trigger the rotation of a camera in case a character is not inside the camera's view. The camera constantly rotates until both characters are again included in the frame. There is another element in this process called Average. The camera targets an invisible object called Average, which is located halfway between the characters. The camera is slowly rotating around Average making sure that all the characters are included in the frame. When one of the characters starts to move, the camera can compute the new position of Average, an invisible object, and rotates at a constant distance from Average.

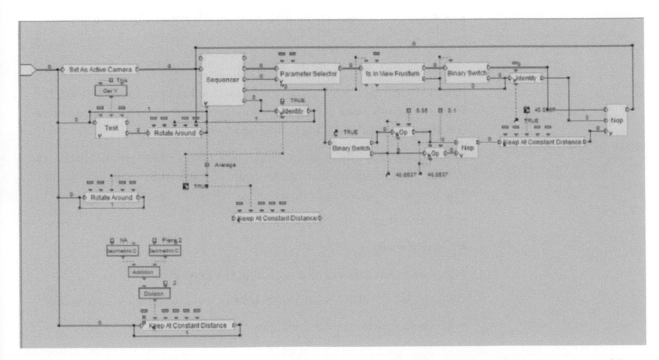

This schematic view of the stabilization behavior shows how several strings of building blocks control various parameters of the camera's behavior. The sequencer on the top row of elements is first testing if the characters are in the Is In View Frustum of the camera and moves the camera to Keep in Constant Distance from the Average invisible object or 3D frame.

3   HOW TO DESIGN BEHAVIORS FOR INTERACTIVE CAMERAS THAT PRODUCE THE SAME EFFECT AS CAMERAS FOUND IN MOVIES

By outlining interactive relationships between the camera and its environment, we can design building blocks that can mimic the cinematographic experience inside a virtual world. The following groups of building blocks can be encapsulated in more complex behaviors that converge into new examples of virtual cameras, such as advanced dialog cameras, that have not yet found their counterpart in the traditional cinematographic experience. The examples covered in this book describe three types of virtual cameras including the following:

- The "Hitch" camera, which is an automated observation camera inspired by Alfred Hitchcock's 1954 film *Rear Window*

- A path camera inspired by sequences from Stanley Kubrick's 1957 *Paths of Glory*, 1968's *2001: A Space Odyssey*, and 1980's *The Shining*

- A free-range camera inspired by Antonioni's *L'Avventura* in 1960

3.1   Notes about *Rear Window* by Alfred Hitchcock

Alfred Hitchcock creates a unique relationship between James Stewart, the viewer/voyeur looking at people across a courtyard, and scenes taking place across a courtyard in the back of an apartment building.

By shooting the whole movie from an exclusive viewpoint, Hitchcock invents a Hitch camera that can be easily recognized by its unique feeling. The parameters of such a camera are as follows:

- Shooting from a fixed point of view

- Using a zoom

- Having restricted field of view across the courtyard

- Being able to follow people through the windows of their apartments

- Shooting constantly night and day

3.2   The Design of the Hitch Camera

Let's explore the design of a virtual camera using the same *Rear Window* parameters. The Hitch camera is looking for characters who are members

In the movie *Rear Window* Alfred Hitchcock creates a unique relationship between James Stewart playing the viewer/voyeur looking at people across a courtyard and scenes taking place across the courtyard in the back of an apartment building.

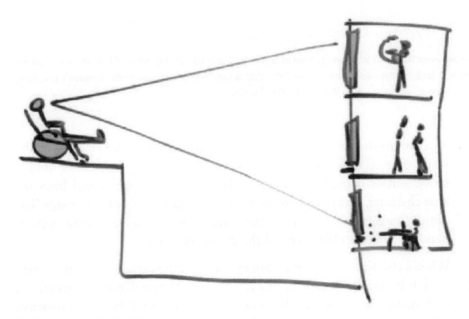

Simple sketch of the "Hitch" camera setup.

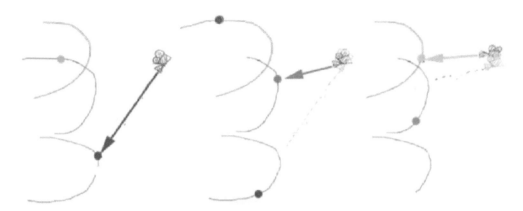

When a character shows up within the range of the camera, it becomes the camera's target.

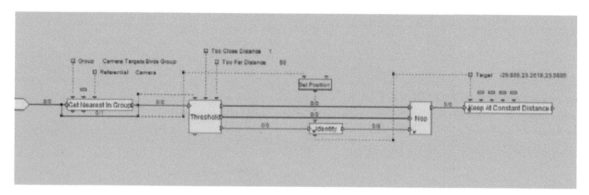

This schematic view of the Hitch behavior illustrates the management of a group of targets—characters in this case—who belong to the same group. The camera will look for the nearest characters from that group that are closest to the camera's proximity as they appear within a certain range or distance set by the threshold behavior.

of a group—for example, the tenants of the building showing up at the window or on their balcony. The camera targets the red character until it moves out of range. The camera looks for a new character and finds the blue character, which happens to be entering inside the camera's range. The camera targets the blue character, and then the same process repeats between the blue character and the green character.

When a character shows up within the range of the camera across the courtyard, it becomes the camera's target. The interactive demo for this virtual camera can be found on the companion CD-ROM in the Interactive Camera folder under FollowCamChasebird.cmo.

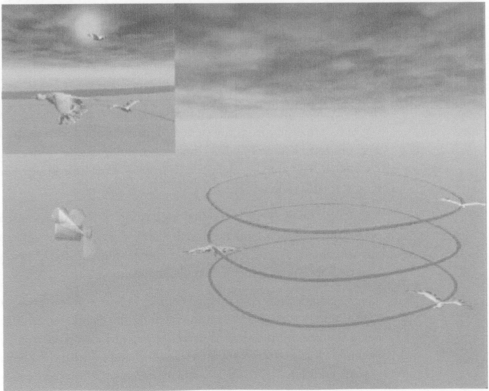

Sequence of frames from a Hitch camera targeting birds. The Hitch camera is designed after the *Rear Window* camera setup *(continued)*.

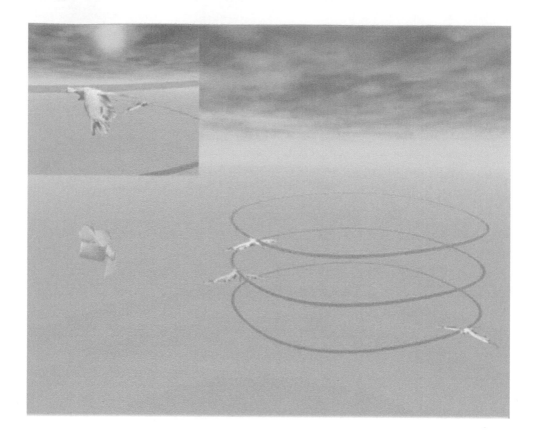

This virtual camera can automatically choose to film a flying bird according to its proximity from the camera. This behavior is also useful for shooting selected characters while they are moving inside a virtual crowd of characters. The camera identifies characters from this group who are the closest to the camera.

The following example is another application of the Hitch camera concept where the camera is guided through a crowd. The interactive demo for this virtual camera can be found on the companion CD-ROM in the Interactive Camera folder under CocktailCamera.cmo.

This schematic view of the Hitch camera behavior shows how the location of the camera is dependent on the position of a character. First, the camera using the proximity sensor detects the nearest character from the group. Second, it locks its target on the targeted character with the Look At behavior. An additional rotation constraint added to the green cameras keeps

Sequence of frames from several Hitch cameras looking at characters. The green cameras are targeting characters as they walk nearby *(continued)*.

Sequence of frames from several Hitch cameras looking at characters. The green cameras are targeting characters as they walk nearby.

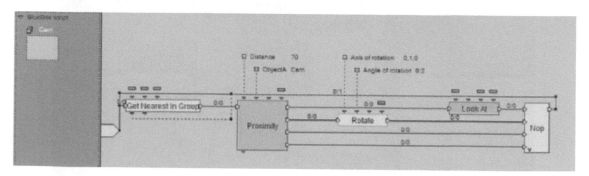

Schematic view of the Hitch camera behavior.

them searching and prevents them from staying too long with one character, therefore creating the illusion of moving through a crowd.

3.3 Notes about *Paths of Glory* by Stanley Kubrick

A few movie examples can help us while designing interactive cameras that emulate a film director's way of looking at the world. Kubrick films the trenches of war in *Paths of Glory*, an astronaut jogs in zero-gravity corridors of a spaceship in *2001: A Space Odyssey*, and a little boy runs through a maze in *The Shining*. World War I trenches, corridors in a spaceship, and a maze are Kubrick's variations of a similar path camera motion. Although the types of cameras available have radically changed between the use of standard dolly motion for the *Paths of Glory* shots in 1957 and the Steady-cam used for the maze scene in 1980's *The Shining*, Kubrick's conceptual design remains the same, with a moving camera chasing a character inside a walled environment. The camera is not alone; it is in a relationship with groups of elements that are part of the scene.

In *Paths of Glory*, Kubrick creates a dynamic relationship between the camera following an officer walking in the trenches and soldiers moving to the side to clear the way in front of him.

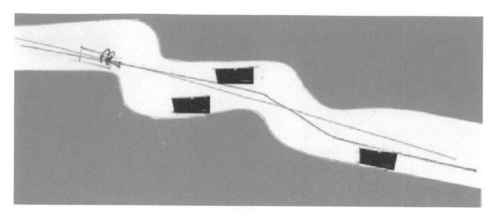

The camera meanders inside a trench to avoid soldiers and other obstacles.

In *Paths of Glory*, the viewer feels as if the environment affects the motion of the camera. Kubrick creates a dynamic relationship between the camera following an officer walking in the trenches and soldiers moving to the side to clear the way in front of him. The path camera following the officer interacts with the soldiers clearing the way for the officer walking in front of the camera. When the camera finds its way between soldiers walking in the trenches, the camera can also change according to the physical attributes found in the environment. The camera meanders inside the trench to avoid obstacles.

The scene can be broken into two groups of elements interacting directly with the camera:

• Soldiers and obstacles affecting the camera's movements

• Collision detection and sliding with friction on the walls of the trench

When the attack starts, Kirk Douglas, the officer, climbs the wall of the trench and runs through the open battlefield. The contrast with the previous scene is striking. A dolly camera follows closely the officer running with his battalion through the hellish battlefield. The trench seen through the path camera was already scary but nothing compared with the frightening battlefield where most of the people run to get killed.

3.4   The Design of a Maze Camera

Let's study these scenes to recreate a virtual camera with interactive behaviors that can mimic Kubrick's camera inside a virtual set. Virtual cameras

Inside the trench.

can also avoid characters and change direction when getting too close to obstacles.

We can duplicate this effect by building a close relationship between a moving camera and characters acknowledging the viewer/camera. Opening a path in front of the viewer while moving inside a crowd of characters can create stunning immersive effects for the viewer.

In *The Shining*, Kubrick films a chase between a child and an adult in a snowy maze, shot from the point of view of Jack Nicholson's character. The motion of the camera feels to be almost slowed down by the snow. Both the camera and the character played by Jack Nicholson gradually lose ground while running after the boy.

In this example, the surrounding environment of the maze changes while running faster inside the maze. We can create a virtual camera that changes our perception of a virtual landscape according to our speed of motion. The virtual camera will interactively increase speed following the viewer's input.

The attack. Transition from a path camera located inside the trench to a free camera outside the trench.

The landscape will scroll faster and the virtual camera's depth of field will also change, creating an effect of compression in space and time. In other cases, the virtual cameras in motion can change their depth of field in tunnel-like environments when surrounded with too many obstacles or characters. You can find a step-by-step tutorial about the making of a path camera in the tutorial section in Chapter 5. The interactive demo for this virtual camera can be found on the companion CD-ROM in the Interactive Camera folder under DonutDynamicTextureFOV.cmo.

The maze scene is a good example of a camera designed to represent someone walking and looking around in space. The camera can imitate

Running inside the maze.

Step-by-step tutorial on how to make a path camera inside a donut can be found in Chapter 5.

moments of surprise when someone bumps into someone else. It can automatically target characters as they walk close to the camera, then cause the camera to unlock itself and stare at another character.

The Maze camera is a virtual camera mounted on a hinge system with articulated arms, which provide precise motion. Each arm is mounted on springs and articulated with hinges. One arm is attached on one side to the feet of the selected character and the other arm is attached to a vertical column supporting the camera. The distances of the arms of the hinge are set. The interactive demo for this virtual camera can be found on the companion CD-ROM in the Interactive Camera folder under CocktailCamera.cmo.

The architecture of the behavior of the Maze camera is divided between two groups of behaviors:

- The Constraints group covers collision detection and keyboard controls. Constraints behaviors include image stabilization and motion using the hinge system.

- The Character Follow group includes the constraints defining the cocktail camera that allow the viewer to look around at other characters. The

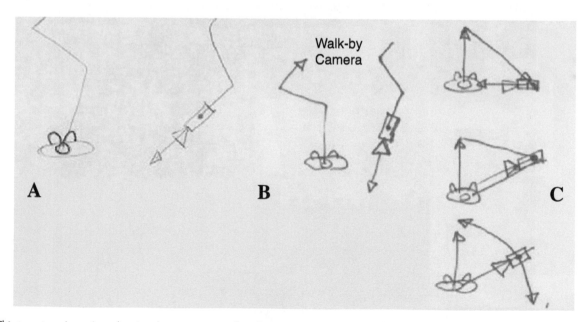

This top view shows how the virtual camera can pick a character, (A) and (B), and the hinge system once the camera is attached to the character.

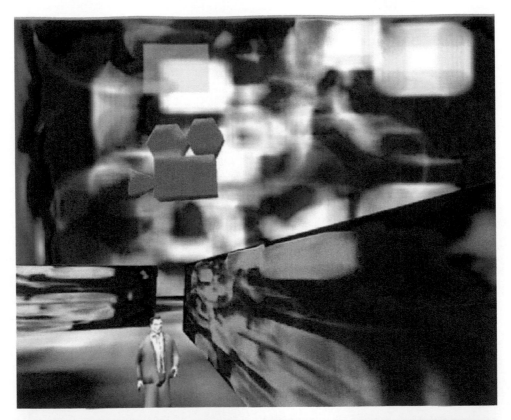

This sequence of four frames from a Maze camera shows how the camera targets characters as they walk closely to the camera.

This sequence of four frames from a Maze camera shows how the camera targets characters as they walk closely to the camera.

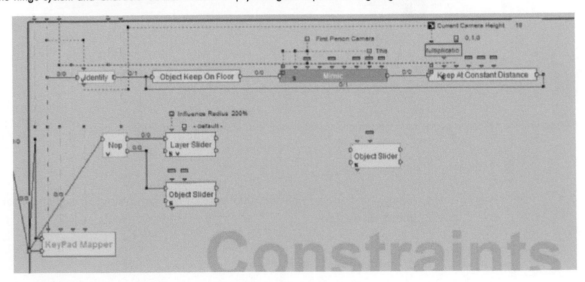

The schematic view of the chase camera shows the architecture of the two main behaviors. Constraints covers stabilization with the hinge system and Character Follow covers the psychological aspects of targeting characters.

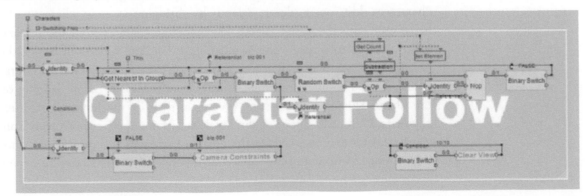

The Constraints group covers collision detection and keyboard controls.

The Character Follow group includes the constraints defining the cocktail camera that allow the viewer to look around at other characters.

In *L'Avventura*, the helicopter comes from the sea and makes a large turn above the island before landing.

Character Follow behaviors take care of the psychological aspects of targeting characters.

3.5   Notes about *L'Avventura* from Antonioni

In *L'Avventura*, Antonioni builds a visible psychological tension between characters lost on a desert island by bringing a new character by helicopter onto the desert island. The helicopter comes from the sea and makes a large turn above the island before landing. Antonioni brings the helicopter as a new character into the scene that can accelerate the course of the story. The flying machine follows a perfect path before landing on the island. This turn of the story creates a "deus ex machina" effect, literally using God walking out of the machine, a known effect borrowed from classic theater and reserved for the entrance of kings and princes performing on stage. Antonioni reinforces the presence of the machine by having all the characters look at the helicopter while landing. The tension of the scene is relayed to the viewer by using a camera positioned at eye level and tracking eyes following motions of surrounding characters.

Antonioni develops several experiments with cameras in the desert island scenes of *L'Avventura*. Antonioni's camera sweeps across the desolated landscape with the insistence of a robotic camera. The camera seems to scrutinize the rocks looking for clues about the missing character, waiting for something that may come out of nowhere. The camera orbits around groups of actors with an autonomy of movement that goes beyond the recording of a drama taking place between the characters. Antonioni's camera is looking for a missing person with more determination than some of the characters wandering around the island. The camera seems to have a life of its own and to be on a mission to find the missing character.

Antonioni suggests that a camera can also have a physical presence, with its own autonomy and self-determination. The camera frames landscape and people in almost a nonhuman way. For example, the camera frames rocks with expressive shapes and actors standing against the sky. The camera looks at the world like an image recognition system trying to recognize a human face inside a puzzle of surfaces. The camera is an autonomous vision system that is more than a floating viewpoint or a viewing/recording machine. The camera's vision is made of topologies of rocks and of shapes of actors' bodies seen against the sky. This autonomous vision of the world could be called the architecture of vision of Antonioni's camera.

Antonioni, an architect and filmmaker, suggests the possibility of creating a camera with a presence, an autonomous language, and a vision independent of the story being told. We can experience a similar process with architecture. Architects can design nonfunctional elements that have a fundamental physical presence in a place. These autonomous architectural elements—for example, volumes casting games of lights and shadows—exist beyond the functional needs of the people living inside the building.

A virtual camera inspired by *L'Avventura* could achieve a total convergence between camera techniques, physical presence of the camera, and viewer's psychology. This new virtual camera would have a tangible presence with self-determination and autonomy of decision. But *L'Avventura*, a virtual camera expressing forms, color, and time through a new architecture of vision, remains to be designed.

4   CONCLUSION

We have seen in this chapter several examples of advanced cameras that could go beyond the simple framing and recording of a scene. Advanced cameras can be complex alliances that can define their own presence and reshape space and time. Kubrick combines vision and motion to create a vortex effect that seems to swallow the space of the world around the edges of the frame. The camera filming the scene inside the maze in *The Shining* is an example of Kubrick's vortex camera; the curved space around the running boy seems to be absorbed by the camera. Similar deformations of space can be found in the last part of the movie *2001: A Space Odyssey*.

Alfred Hitchcock's *Rear Window* explores a man/machine alliance combining an actor, James Stewart, a camera with a zoom lens, and a wheelchair. The result is a brilliant distortion of time inside a place that cannot be changed. The last example is taken from Antonioni's *L'Avventura*. Antonioni creates a vision machine that is ready to take over our own system of perception. Mapping perceptions seems to be the final frontier of virtual cameras and raises exciting issues of representation. Can we explain how to represent our dreams? How can autonomous cameras represent their own dreams? Whose dreams are they: yours or theirs?

Filming and writing have shared the same ability to communicate what has happened in a scene. Writing and sketching are instant media in which the author is the viewer of the content being created. A writer writes and rewrites a story, but the film director shoots and edits. What happens when film director, editor, and viewer are the same person? A movie camera can be conceptualized as an observer floating in space in time. Although there is no need to process my page to read it again, I need to stop writing to read it. Sometimes holding the pen, typing, and formulating sentences work together in a seamless musical harmony. Other times the writing process is discontinuous and often interrupted. The timing of the real-time interaction between writer and reader is similar to the virtual camera holding the viewer's interest in a virtual space. Fragments of reality or fantasy worlds captured by the camera are influenced by various attributes of the set that can include lighting, motion, and colors and some of the camera's attributes such as focus, depth of field, frame rate, and movement.

In comparison to writing, cinema is young—about 150 years old. Filmmakers constantly reinvent relationships between the camera and content

captured by the camera. To design a virtual camera, we can try to understand what links the cameraman's activities of shooting and framing a scene
and the final experience of the viewer watching the camera work. The
viewer's experience taking place in a delicate area somewhere between the
eyes and the brain is one of the highest expressions of the viewer's freedom,
independent from the technology used. Compelling content can be
revealed by low-budget productions the same way that noble ideas and rich
scenes can be wasted in state-of-the-art but disappointing movies.

The sequence of actions that leads to the making of a movie can take several
months, and in some cases years, from the time the shooting begins to the
final cut. It goes down to about a 60th of a second in the case of a virtual
camera. The cameras requested for these different media are also using
completely different spaces. Cinema and video require physically present
cameras inside the physical space. Although cinema, video, and virtual
cameras have little in common, they can communicate to a viewer through
an old concept called the frame. *Pursuit of Happiness: The Hollywood
Comedy of Remarriage*, Stanley Clavell's essay on his own moviegoer experience, explains the role of the frame as a space- and time-defining element
of filmmaking. He writes from the perspective of moviegoers sitting in the
darkness of a movie theater and contemplating the big screen. He describes
poetic associations that take place between what is seen on the screen and
what is unseen beyond the edges of the frame—unseen, unshown, but sug-
gested.

Stanley Clavell's study of the viewer's experience explains how we can
imagine the continuity of the space from seeing isolated fragments of visual
content inside frames. He describes the back-and-forth movement that
takes place between fragments—what we see inside the frames of the movie
and the imaginative world of the movie that we recreate in our mind.
Hitchcock's *Rear Window* is built on our fragmented perception of the life
of a whole building. James Stewart, who cannot leave his wheelchair, has
plenty of time to shoot pictures of the occupants of a building seen across
a courtyard, but when he falls asleep we miss crucial cues about the timing
of a murder. The plot is clearly constructed around gaps in space—we do
not see the whole picture of the building or we do not see other viewpoints
that could reveal more information about the murder. Hitchcock shows the
impossibility of recording the routine of everyday life in all its detail and
the unpredictability of even well-understood routine. Hitchcock's poetic
meditations about the mystery of the routine of everyday life are very

helpful to conceptualize unpredictable events inside a virtual space. Gaps remain something difficult to recreate in virtual spaces.

Real-time content implies that the film director shoots, edits, and screens as the spectator views or that the author writes and publishes at the same time the reader reads it. This ultimate convergence of creation, production, and diffusion has already been experienced in the first movies. At the end of the nineteenth century, the Lumiére brothers created a cinematographic environment where an operator could shoot, process, and project a film in the same day. The order of appearance of events is still totally controlled by the film director or the author of a book. Real-time content puts new constraints on the narrative that needs to be multiple, interactive, and filled with dramatic elements or nodes. The traditional hierarchy of authorship is decentralized and shared between spectators/users who can interact with the story process and physically share the time and space of the narrative— Spectators become users/viewers/spectators sharing an immersive experience, which is somehow believable. Virtual reality users can experience too much freedom, the possible source of problems such as the loss of control over the camera. The new degree of freedom that comes with this medium has a great effect on the design of virtual cameras inside virtual worlds.

CHAPTER 7

Advanced Virtual Cameras
and Their Environments

 This chapter presents three types of advanced virtual cameras created for three different virtual world projects. A copy of each virtual world can be found on the companion CD-ROM. For each project, the production process is described step by step and illustrated with screen captures of the files available under Chapters 6 and 7 on the companion CD-ROM.

1   CONCEPTUAL DESIGN

The description of the virtual worlds will cover early conceptual design, the building of a virtual world, and the implementation of a virtual camera inside a virtual set. The virtual worlds described in this chapter are the diner scene, the Nighthawks installation, and the Aphrodisias project.

The description of the virtual worlds can be summarized with a checklist of critical questions for the creation of a virtual world.

- What is your storyline?

- What happens when you start to play inside the virtual world?

- What is involved in the preparation of your project?

- How many types of input devices do you provide to the viewer?

- Is there an automated mode for virtual cameras and characters?

- Do you provide multiple views or additional views?

You or your team can use the checklist in the early planning stages of the design process or during production to ensure that you control all the aspects related to the design of your virtual project. The checklist can help to maximize the production effort on critical topics without spending

The diner scene is an interactive dialog between virtual actors in a bar.

Nighthawks is an interactive installation inspired by Edward Hopper's painting.

significant time on nonsignificant elements or secondary topics. The checklist also may be helpful to reduce time spent on redundant elements during the production process.

Although the examples of virtual worlds presented in this chapter are set in different contexts and were created for specific audiences, they share similarities in the way they were produced. The following lists some of the common features shared by the projects presented in this chapter:

- The projects use advanced virtual cameras that can be either fully automated or user controlled. The design of advanced virtual cameras uses the concepts and behaviors illustrated and analyzed in Chapter 6.

- In the opening scene of each virtual world project is an autonomous environment that can generate its own content without input from the viewer.

- The virtual worlds can also react to the viewer's presence. Viewers can influence the virtual world rather than take control of the action in the world. For example, virtual characters acknowledge the presence of the viewer in the Aphrodisias project.

- Because they are mostly event driven, characters and cameras never repeat the same paths and use only a few canned animations.

The Aphrodisias project is a simulation of the life on a Greek agora.

The three-step checklist includes the conceptual design, cameras, and navigation. The viewer's experience covers many elements found inside a virtual world.

A project starts with conceptual design ideas that are often hand sketched on paper. During this "conceptual phase," we try to define a viewer's experience rather than a precise project. We also look at similar existing projects to understand the specificity of what we are trying to achieve.

1.1   Step 1: Conceptual Design

Answering the following questions will help to give you a clearer idea about your project.

1 Write an introduction to your project that includes, but is not limited to, the following: What are the general context, time, and location of the project? How would you describe someone's experience with the virtual space that you are creating? Is the concept related to artwork, to scientific simulation, or to entertainment or gaming?

2 Description of the project. What is the storyline? What happens when you start to play inside the virtual world?

3 Project preparation and construction. What is involved in the preparation of your project? Please check the workflow handout described in Chapter 2 for a list of your tasks. Where do you anticipate limitations or difficulties?

4 Questions related to the interactions between visitors inside the virtual space. How many types of input devices do you provide to the viewer? Please describe each interaction using a keyboard equivalent function. How many types of interactions are connected to the viewer's input devices? Describe in detail the interactions. List the input devices, the length of the interaction, how each interaction starts, and where it ends.

5 Advanced automated functions and artificial intelligence systems inside the virtual world. Is there an automated mode for virtual cameras and characters? Are they autonomous? Are they target/goal driven? Can you describe basic interactions between characters and between virtual cameras and characters? Do you provide specific interaction between the camera–viewer duo and characters. Do you have ways for characters to acknowledge the camera?

6 Display. Do you provide multiple views or additional views? What is the order of the sequence between multiple views? What is the relationship between additional views? Do you provide a hybrid mode that can switch the commands of the virtual world between viewer controls and automated mode?

1.2   Step 2: Cameras and Navigation

In Step 1 we addressed the conceptual design. In Step 2 two we look at cameras and the ways cameras can navigate inside the virtual world.

Please answer the following questions about cameras and navigation (the questions and answers may summarize previous discussions about the project you may have had during Step 1).

1 Questions about cameras. How many scenes do you have? List and name the cameras needed for each scene of your project. Describe the main attribute for each virtual camera. Is the virtual camera one of the following?

- Follow camera

- Over-the-shoulder camera

- First-person camera

- Third-person camera

- Dialog camera

- Transitional camera

- Dolly camera

- Orbital camera

- Chopper camera

- Jet camera

- Car camera

- Other camera

Is the first virtual camera used at the opening of the scene a path camera or a free-range camera?

For each camera, do the following:

- Describe the content seen by the viewer inside the frame.

- List and rank the cameras in order of priorities for the scene to be best understood.

- List and rank the cameras in order of appearance for the viewer.

2 Questions about navigation. Is the interactive navigation controlled by the user, by an automated system, or by an artificial intelligence system? Describe the input device used for controlling the virtual camera. Is it one of the following?

- Mouse

- Keyboard

- Game controller

- Phone

- Sound input

- P5 glove

- Serial input device using a serial extra in Macromedia Director

- USB input device using an extra in Macromedia Director

- Other input device

Describe additional interactive parameters—for example, change of speed, change of color, translate, rotate, scale, character animation controls, show and hide, and switch script.

1.3   Step 3: The Following Questions Are Related to the Viewer's Experience

1 What can the viewer explore?

2 What are the options given to the viewer?

2   ADVANCED VIRTUAL CAMERA DEVELOPMENT FOR THE DINER PROJECT

Created by Jean-Marc Gauthier with the help of Fabien Barati and Zach Rosen

The Diner project is inspired by a conversation between two characters inside a small bar. The movement of the virtual camera, called the Dialogue camera, follows the intonations of the viewer's voice or sound level in the physical space of the installation. The following conceptual process evolved from notes from a discussion between Fabien Barati and me during the design process of the Dialogue camera for the diner. The interactive demo for this virtual camera can be found on the companion CD-ROM in the Virtual Cameras folder under Dinner15z.cmo.

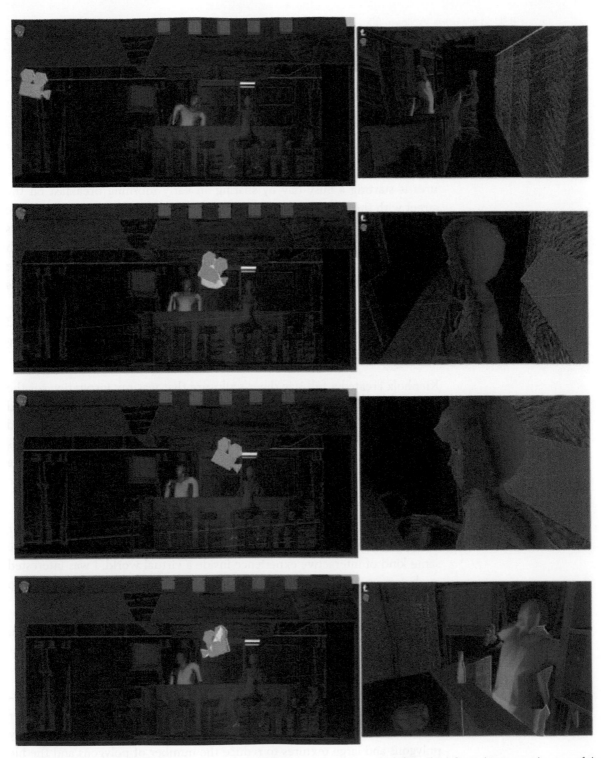

A sequence of one movement of the Dialogue camera from one character to the other. The left window is a side view of the diner with the virtual camera—in green—moving through the space. The right window is the viewpoint of the virtual camera. The movement of the virtual camera is activated by the viewer's voice or by the sound level in the physical space of the installation.

You can open the demo for the diner project, found on the companion CD-ROM, to experience the features described in the following checklist.

2.1   Step 1: Diner: Conceptual Design

1 Starting point. When traveling to a new city, one of my greatest pleasures is starting a busy day by sharing an early morning breakfast on the street with the local people. Although the ingredients are almost always the same, there are numerous ways to get a breakfast in the street with endless new surprises. I remember the greasy omelettes from Washington; eggs with mayonnaise from Paris; and, of course, Manhattan Greek diners, which are on top of my list. The busy mood of these early morning breakfasts reminds me of Dore's scenes of urban congestion in nineteenth-century London. Edgar Allen Poe's fantastic novels describe a typical pedestrian who would spend day and night wondering through his own city, following and eating with crowds wherever they are at any time of the day and the night.

Kienholz created a full-size diner sculpture that was my inspiration for the atmosphere created in the virtual diner. The work of Kienholz is influenced by surrealism, hyperrealism, and pop art installations. His sculptures and installations convey the feeling of the atmosphere—a scene that goes beyond the representation of people and their activity. His installation is a real diner recreated inside a box that you can experience from the inside and the outside.

2 Description of the project. The scene that takes place inside a diner is visible from multiple viewpoints. This gave me the idea of creating the same kind of interactive experience inside a virtual world. I was interested in designing a virtual camera that can use motion in the diner scene to convey the sense of multiple actions happening at the same time.

The virtual camera can react to people's voices and sounds inside a room. For example, viewers can interact with the virtual camera by talking or clapping their hands.

3 Project preparation and construction. The goal was to create an installation that would retain the complexity of the atmosphere of a diner. After modeling the diner in Lightwave, the 3D model is stripped of unnecessary polygons and large textures to reduce the number of polygons and the file size of textures, and therefore the overall file size of the model. All textures are saved in .jpeg format. In the case of double-sided polygons, the outside

Drawings for the diner project that were inspired by Kienholz' installation.

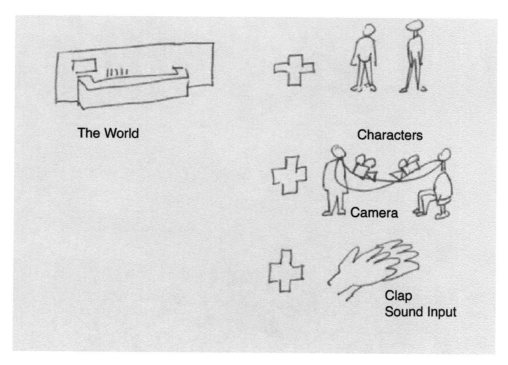

The Dialogue camera can react to people's voices inside a room. Viewers can interact with the Dialog camera by clapping their hands.

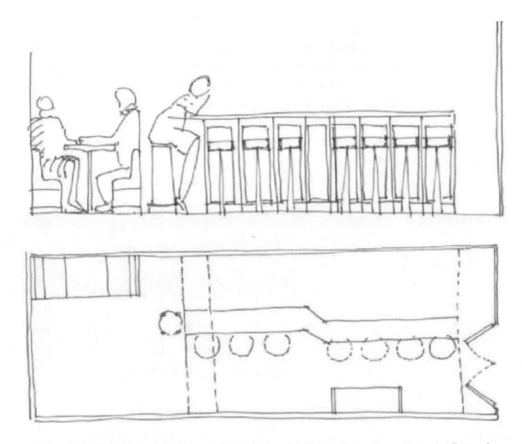

Blueprint of the virtual world used as a template to build the 3D model. The goal was to create an installation that would retain the complexity of the atmosphere of the diner.

face of the polygon is deleted. As a result the inside of the box can be seen from cameras located outside the box, which results in what is called a culling effect. This process gives more flexibility for camera placement and cuts down the number of polygons. Polygon reduction and texture compression help to spare rendering power needed for animated characters. In this case, the final 3D model has 5,000 polygons and only three textures.

The female character is modeled in Lightwave. The modeling of the character's head requires more details because the virtual camera will frame the head with close-up shots. Low-resolution modeling techniques are applied to parts of the body that will be filmed from far away. The interruption of the modeling of the body in parts that are covered with clothing helps to reduce the number of polygons even more. The final low-resolution model shows 3,786 polygons, which is acceptable if the viewer is using a 3D acceleration card.

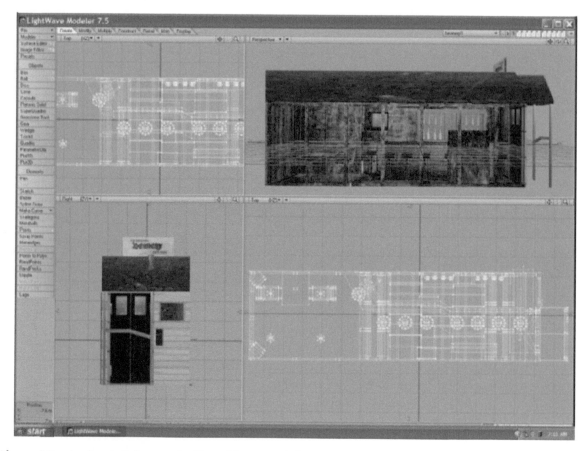

After modeling the diner in Lightwave, the 3D model is stripped of unnecessary polygons and large textures to reduce the number of polygons and the file size of textures.

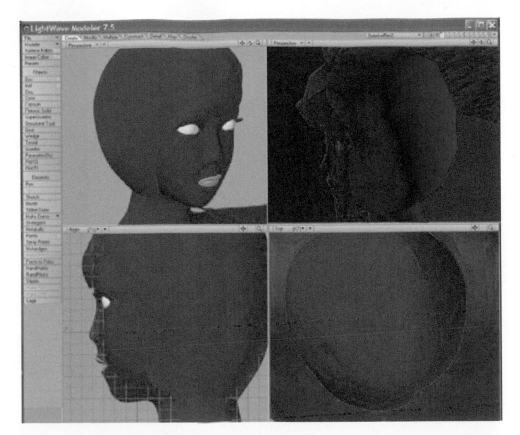

The character's head requires more detail because the virtual camera will frame the head in close-up shots.

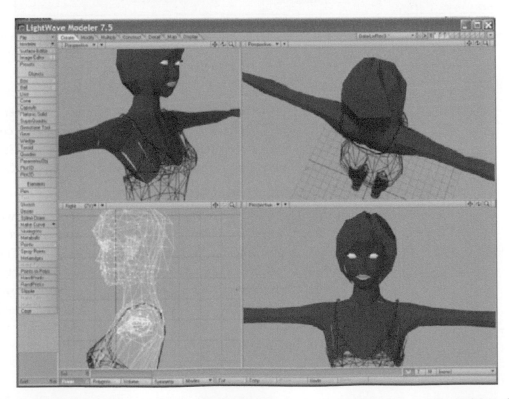

Low-resolution modeling techniques are applied to parts of the body that are filmed from a distance. The modeling of the body is interrupted in parts covered with clothing, which helps to reduce the number of polygons.

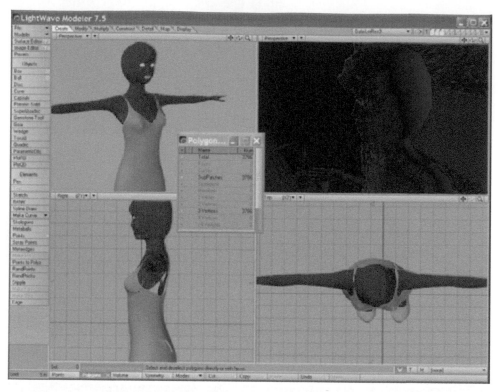

The final low-resolution model shows 3,786 polygons, which is acceptable if the viewer is using a 3D acceleration card.

The female and male characters are animated in 3Dmax Character Studio, taking advantage of Character Studio's biped. You can find a complete description of animating 3D interactive characters in Jean-Marc Gauthier's book *Creating 3D Interactive Actors and Their Worlds,* by Morgan Kaufmann Publishing.

Although the 3D models are relatively simple, they are loaded with long animations that increase the file size of the virtual world. For example, a virtual world of the diner with characters is about 2 MB. The same virtual world is 20 MB when animations are added to the characters. This problem comes from using canned animation. These well key-framed sequences of gestures tend to overload a project looking for speed, flexibility, and responsiveness. Event-driven character animation, covered in Chapter 8, will help you to design character animation without key-framed animations.

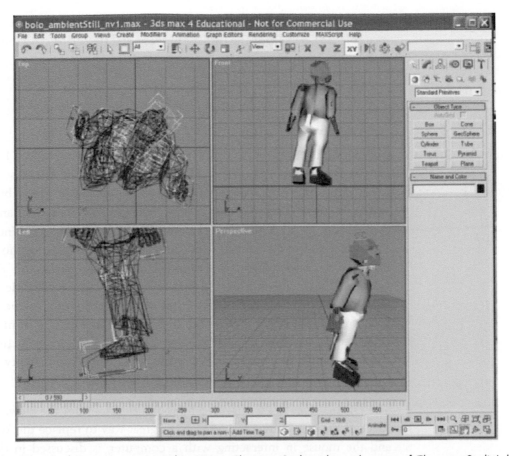

The female and male characters are animated in 3Dmax Character Studio, taking advantage of Character Studio's biped.

4 Interactions with visitors inside the virtual space. This installation explores the possibilities of using several large screens to broadcast the points of view of several interactive virtual cameras inside a scene. The installation is designed for people moving through urban spaces where they are unlikely to stop to use a kiosk with a touch screen, a mouse, or a keyboard. The constraints of urban space limit us to a few input devices involving wireless connections such as PDAs (personal digital assistants), cellular phones, or simple invisible sensors such as hidden microphones that can respond to gestures or sound.

This particular installation is using such a microphone as a sensor hidden behind the screen. The audience can guide the virtual camera by clapping their hands, talking, tapping their feet, or snapping their fingers. The repetition of the same sound can also trigger different motions—for example, one clap sends the camera forward and two claps send the camera backward.

2.2   Step 2: Cameras and Ways to Navigate Inside the Virtual World

1 Designing a new virtual camera called Dialogue camera. The Dialogue camera uses a stabilization system that is similar to the one described in Chapter 6. Once the camera is attached to a character, the stabilization system keeps the camera stable around its target. We can detach an orbital camera from a character—the camera's center of rotation—by increasing the distance between character and camera. At a given distance the camera moves out of the character's range. The camera breaks the Keep at Constant Distance behavior and becomes a free camera. The viewer can also choose to detach the camera from a character by pulling the camera away from the character. We chose to pull the camera away from a character because we thought that pushing a camera inside a character was a counterintuitive way to move a camera. We created a possibility for the camera to slide when getting close to the character.

The Dialogue camera is triggered by sound coming from the viewer or the space of the installation. Sound input, a great way to replace the keyboard and the mouse in interacting with a computer, is discussed in detail in Chapter 8. The volume of sound coming from the sound sensor, a microphone, is tested for values; above a given threshold, the Test building block triggers a message telling the camera to change its target. This is the starting point of a sequence of adjustments in space that will take the camera from one character to the other. The following diagrams illustrate the architecture of the behaviors created for the Dialogue camera.

The Dialogue camera behavior receives a message from the sound sensor and activates two strings of behaviors. The string on the top controls the translation and orientation components of the camera's motion in space. The string at the bottom picks up a new target, which is the other character.

The camera design can respond to many situations that may arise randomly during a viewer's interaction with the scene. The following list covers various responses to situations involving the virtual camera:

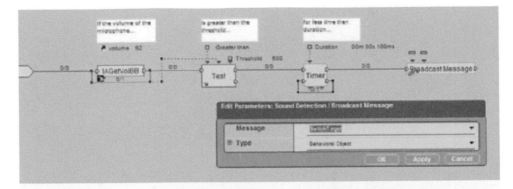

The volume of sound coming from the microphone is tested. Above a given threshold, the Test building block triggers a message telling the camera to change its target.

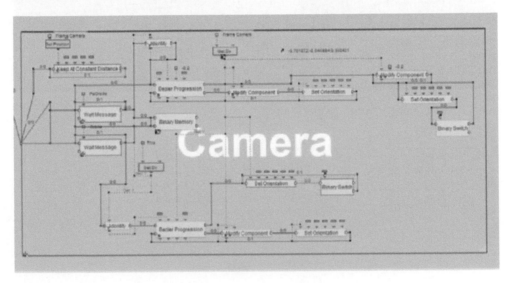

The Dialogue camera receives a message from the sound sensor and activates two strings of behaviors.

- The camera can collide with characters without pushing the characters.

- The camera can check the distance between the character and itself and then choose a character according to his or her proximity to the camera.

- When looking at a character, the camera moves until it targets the character's head. After being attached to a character, the camera can still turn around the character. The camera can be detached from the character and attracted again after 2 seconds if the camera remains close to the character.

Character viewed from the Dialogue Camera in orbital mode.

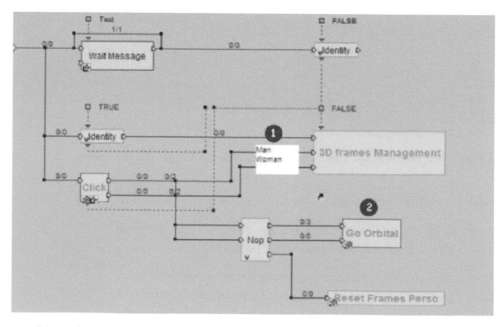

Schematic view of the Dialogue camera's behavior. Step 1 shows how the camera moves across the room to target a specific character. Step 2 shows how the Go Orbital behavior is activated when the camera gets close to the character.

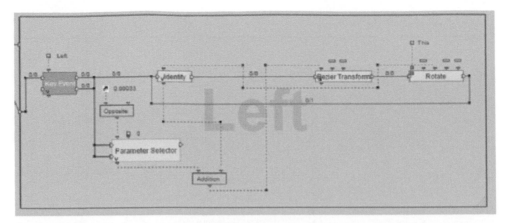

Schematic view of an intermediary stage of the design of the Dialogue camera. The keyboard controls movements of the virtual camera across the space of the virtual world.

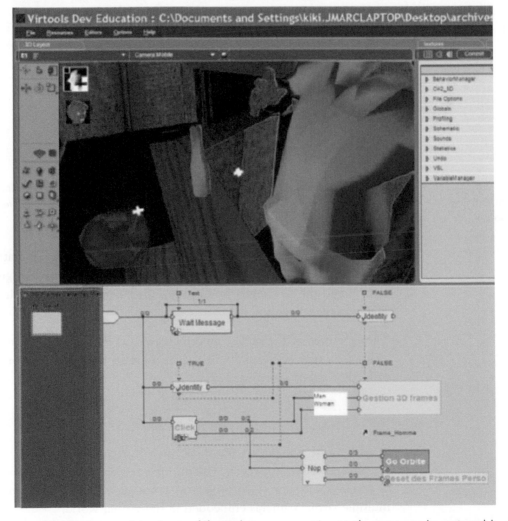

Schematic view of the final stage of the design of the Dialogue camera. The virtual camera can be activated by the sound sensor and also receive other inputs to move across the space of the virtual world.

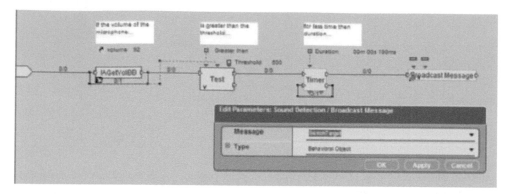

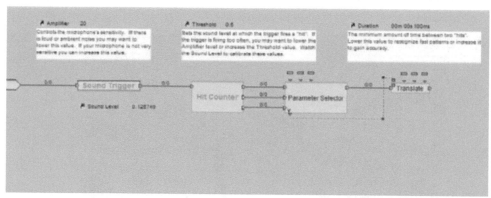

Schematic illustrations showing the evolution of the Sound Analyzer behavior from detecting one sound to detecting a sound repeated several times. On the top, a single clap of the hands can move the camera forward. On the bottom, the repetition of the same sound—for example, several claps of the hands—moves the camera backward.

2 Navigation. As mentioned earlier in this chapter, this installation uses a microphone built into the floor as a sensor. The audience can guide the camera and the characters by clapping their hands, tapping their feet, or snapping their fingers.

The repetition of the same sound increases the duration of the input—for example, a repeated sound will trigger a longer backward motion, similar to holding down a keyboard key.

2.3   Step 3: The Viewer's Experience

In addition to the sound interface, the project is displayed on multiple screens and multiple cameras. The multiple screen technique allows the use of several screens or video projectors receiving video signals coming from multiple virtual cameras embedded in the virtual world.

Lighting interactive characters inside the scene.

The final design was rebuilt from scratch following the lessons from the previous phases. The virtual camera seems to perform differently according to the file size of the animated characters. Please open the copy of the diner on your CD-ROM and go to the Schematic window. You will find the variables for the virtual camera grouped in a control panel so you can tune up the rendering speed of the scene without interfering with the camera's behaviors.

3   VIRTUAL CAMERA DEVELOPMENT FOR NIGHTHAWKS

Created by Jean-Marc Gauthier with the help of Zach Rosen. The interactive demo for this virtual camera can be found on the companion CD-ROM in the Virtual Cameras folder under Nighthawks.vmo.

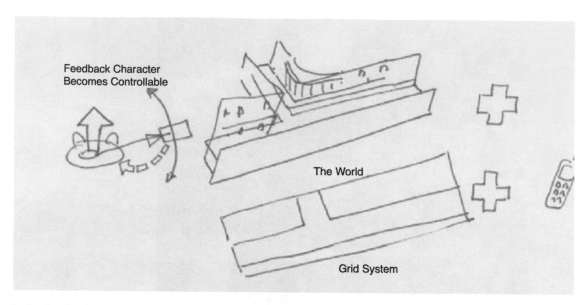

Early sketches for the Nighthawks project show the integration of three elements, from left to right, animated characters with virtual camera, architectural elements, and interface components such as cellular phones.

In the Nighthawks installation the relationships between cameras, characters, and architecture are experienced by the viewer, but they remain invisible.

(continued)

3.1   Step 1: Conceptual Design

1 Starting point. The starting of a new project is sometimes a sentence, a title with keywords giving a definition of the scope. The Nighthawks project shows the integration of three elements: from left to right, animated characters with virtual camera, architectural elements, and interface components such as cellular phones. Some other times it is a phase of exploration of the unknown. The following text includes excerpts from the diary of the making of Nighthawks. I hope that it conveys a mind feeling free to go in many directions, with a desire to explore and discover that will fire up the reader's creative mind.

"The medieval monk who contemplates the fire ravaging the books of the library of his monastery understands that he is about to start something new without any help from the books. I sometimes feel like this monk does—forced to move away from old manuscripts to rediscover the evidence of his shadow on the ground, an ephemeral shaded area helping him to define his presence before any new space is built again. I need to undo something or to let go to build a new space. Because I am always looking to reuse spaces that worked really well in previous projects, I tend to spend time deconstructing previous projects to delete all storylines or references to anything that would distract from the seed of space that holds the project together. I need to prune the project from everything that is not contained into a simple melody of space, into a basic geometry of elements. I cannot wait to start a new project with the same seed and to build freely again. Where can I find a seed of space? How do I know that it is the right one? I found some of them recently in dreams in which spaces seemed to exist before the story unfolded."

2 How to build the virtual set? The starting point of the design of this virtual world is a 3D scene inspired by Nighthawks, Edward Hopper's painting of a New York street corner.

The real-time 3D installation was designed to immerse several players in a virtual world that recalls the streets of New York City from the 1940s. The 3D world from Nighthawks is similar to our everyday life. It takes us into a 3D painting; inside a diner where the last clients share stories with Joe, the bartender; and outside on the street with New York skyscrapers in the background. Each element of the scene has a life of its own, and from time to time these elements move together and interact.

Expressionist movies from the 1920s and film noir provided other sources of inspiration. I remember a German movie from that period in which

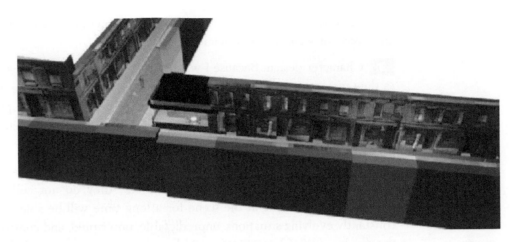

The starting point of the design of the virtual world is Nighthawks, a painting of a New York street corner by Edward Hopper.

buildings with huge bellies extended above the street with air moving between them like the inside of a musical instrument. Fantastic movies have a way of framing buildings the same way we look at people. They notice slight deformations that bring expression to a face, like a new way to look at a facade or a street. I am interested in seeing how expressionist architecture can blur the lines between a built structure and ways to give expressions to a character.

Unfortunately, the 3D modeling tools are not yet spontaneous, physical, or tactile, and they do not offer the feeling of breaking charcoal on paper, scrubbing oil paint on a canvas, or kneading clay with the fingers. The mouse cannot replace the pressure of fingers holding a pencil or pushing clay. When drawing a character with charcoal or graphite, the first lines are fast, intuitive, spontaneous, and precise. When more lines are added to the drawing, too much black threatens to ruin the equilibrium between the white paper and the structure of the forms. Do you experience the same thing while sculpting with polygons? Although the preparation of a virtual set involves the same steps as designing with clay and paper, the time for creativity is a bit different in the digital world.

I focus on relationships between 3D objects, characters, cameras, and environments with an emphasis on the quality and fluidity of the viewer's experience and the optimization of the scene to display high-resolution graphics. I work on the interactive architecture of the project. I like to program relationships between interactive characters and 3D dummy objects (soon to be replaced by 3D models related to the content of the scene). All objects are treated as obstacles, which means that the camera collides with objects and does not walk through them.

3 Character design. Because I want to design characters to look natural as they move through the architecture, I spent quite a bit of time defining the characters' behaviors. Interactive characters acknowledge the presence of the viewer by swaying, shifting, and turning around as the camera moves through the scene.

A group of self-determined characters are able to find their own path across the street. They meet and walk together and they do not collide with objects. People looking at the scene for a long time will be able to follow constantly evolving situations, unpredictable, unscripted, and controlled on the fly by artificial intelligence. Although characters controlled by the viewers display a yellow arrow above their head, they can return to their own decision-making mode when left unattended for more than 30 seconds.

Characters use a set of gestures that we can blend together. Primary animations control the entire body. Secondary animations move only a part of the body. Basic animations such as walking forward, backward, or turning left and right are assigned the standard keyboard control. Other animations can be assigned their own custom messages to be passed on command.

3.2   Step 2: Cameras and Navigation

1 Design of a virtual camera. A first-person camera and a following camera (also called an over-the-shoulder camera) can move like a character and display the world as seen by a character. The camera has two modes: user-controlled and automated.

In the user-controlled mode the numeric keypad of a phone connected to the virtual world can be used to navigate the camera around the virtual set. The keys 4, 6, 7, and 9 rotate the camera left, right, down, and up, respectively. The keys 8, 2, 1, and 3 move the camera forward, backward, to the left, and to the right, respectively. If there is no keyboard or mouse activity after a period of time the camera goes into automated mode until it detects a user pressing keys on the cellular phone keypad. In automated mode the camera follows characters.

The virtual camera constraints include keeping the camera within horizontal and vertical boundaries of the virtual set. The camera is kept on the floor, imitating footsteps and following differences of height between the street pavement and the sidewalks. The camera is also kept inside a grid providing vertical obstacles to prevent it from crossing the facades of the buildings.

2 Optimization of the environment created for this installation. When the project is almost finished, it is time to document behaviors so they can be reused in other projects. Nighthawks gave the opportunity to test new virtual cameras and to develop a system of relationships between cameras and character. Reusing behaviors often means redesigning them from scratch so they can be easily reused in other projects. Teaching is a great way to optimize behavior. There is nothing more gratifying than designing a behavior that happens to be reused by students in different types of content.

Pick up the receiver and press a red key. The yellow arrow selects your character.

In the user-controlled mode, the viewer can control the virtual camera with a phone connected to the virtual world.

The advanced virtual camera behaviors created for Nighthawks include the following:

- Camera Constraints and Stabilization behaviors keep the active camera directly above the 3D frame placed on the ground.

The camera follows the Obstacles level created with the grid placed above the street.

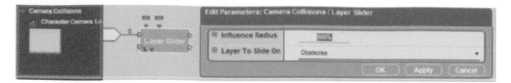

Camera Slider is a combination of keeping the free-range camera inside the virtual world and avoiding obstacles by using Object Slider.

In automated mode the camera follows characters.

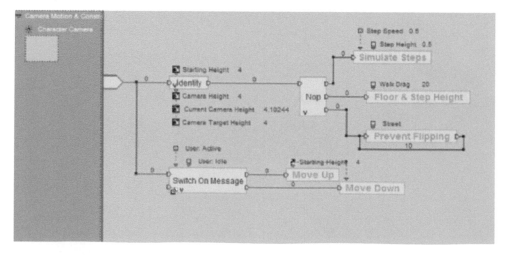

This behavior can simulate the pace of someone walking by raising and lowering the height of the camera.

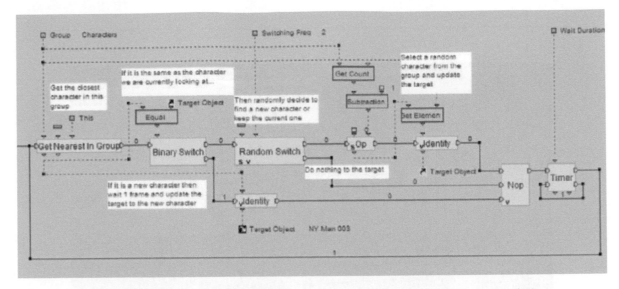

This behavior can be applied to a camera's 3D frame to get the camera to follow characters from a specified group.

- The Simulate Steps behavior can simulate the pace of someone walking by raising and lowering the height of the camera.

- Camera Slider is a combination of a Layer Slider to keep the camera inside the environment and an Object Slider to avoid obstacles.

- Camera Character Selection behavior can be applied to a camera's 3D frame to get the camera to follow characters from a specified group. The behavior follows the closest member of the group first and will then follow anyone who comes closer. You can adjust the frequency with which the camera randomly changes targets.

New character's behaviors include the following:

- Notice Me (action). This behavior should be activated when one character realizes that it has bumped into another character. The behavior sends a message to another character notifying it that it has been bumped.

- Notice You (reaction). This behavior controls how a character responds to being bumped. Possible reactions are a head turning, a step forward, or a piece of dialog.

- Ray Deflection Test. This behavior calculates a deflection angle by firing two adjacent Ray Intersections. The result is a vector that moves the camera away.

Multiscreen display of Nighthawks.

3.3   Step 3: User Experience

Nighthawks gave me the opportunity to rediscover the role played by several interactive media while designing a virtual space. This project is trying to push the envelope by inventing new ways to represent events in a virtual space. It is a place where my interests for 3D animation, virtual reality, artificial intelligence, and gaming can converge. I was not trying to make something new by recycling old ideas but to give a new voice to the scene. The installation inspired by Edward Hopper's famous painting and by movie sets from the 1940s goes far beyond a traditional 3D representation of a scene by adding more cameras and more windows where the scene can be watched at the same time from many angles. Cameras in the scene can move by themselves and make their own determinations.

Nighthawks takes the viewer into a dream of New York in the 1940s. The scenes can also be inspired by our everyday life, including meeting friends at the diner late at night; sharing stories with Joe, the waiter; seeing people outside walking up and down the street; and having New York City somewhere in the background. Each element of the scene has its own personality, and from time to time the elements become alive and interact together.

4   VIRTUAL ARCHEOLOGY AT APHRODISIAS 2003

Created by Jean-Marc Gauthier with the help of Haluk Goksel and Zach Rosen for Christopher Ratte and his team of archaeologists at New York University's Institute of Fine Arts

As described by Christopher Ratte, Virtual Archeology at Aphrodisias is a virtual reality environment that can be experienced as an installation in a variety of settings such as a museum or classroom. It can also be used online as a communication tool by archeologists and students. The virtual environment includes reconstructions of the agora of the antique city of Aphrodisias using related architectural elements, sculptures, and original colors and textures. State-of-the-art, real-time animation and artificial intel-

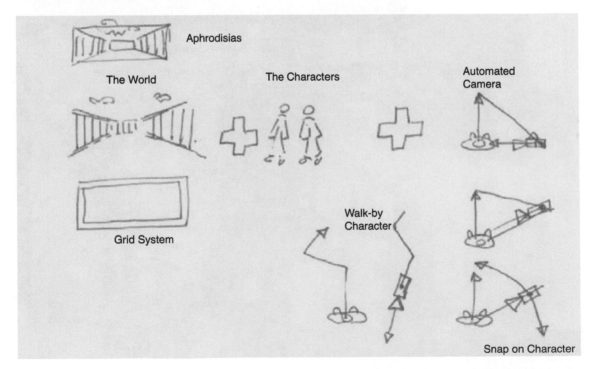

The Aphrodisias project is the integration of three elements shown here from left to right: architectural elements, interactive 3D actors, and an automated virtual camera.

Sketch for the virtual agora.

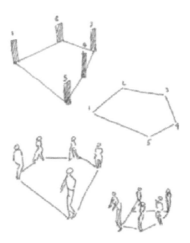

People walking in the marketplace. Where to start? The unknown—May 2002.

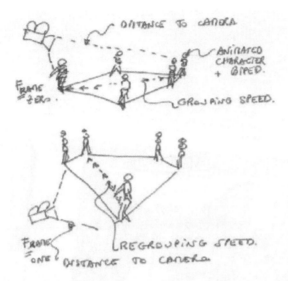

Several hypotheses were made about the possible relationships between a virtual camera and a group of people walking together.

ligence help to recreate believable interactions between the viewers and virtual characters living inside the agora. Thanks to an NYU Curricular Challenge Grant funding research into low-cost virtual reality displays for classrooms, it will not be long before students at the Institute of Fine Arts can learn archeology while walking through a virtual city. The interactive demo for this virtual camera can be found on the companion CD-ROM in the Virtual Cameras folder under Aphrodisias.vmo. Please note that several files with various levels of complexity are available.

A student follows the virtual camera, pointing at horses pulling a chariot inside the agora. View of the portable installation built at ITP-Tisch School of the Arts in 2003.

Aphrodisias is designed as a low-cost immersive installation that can be used by groups of students and teachers to study archeology, history, and urban forms of life.

"Virtual Archeology at Aphrodisias is an architectural model of the agora at Aphrodisias that is populated by a number of self-determined virtual characters," according to Ratte. "The installation differs from a conventional walk-through in the use of artificial intelligence to determine the movements of the figures and the camera, rather than programming them to follow set paths. The result is a closer approximation of a real spatial experience."

A student directs the interactive camera through one of the doors of the agora with an infrared sensor. View of the portable installation built at ITP-Tisch School of the Arts in 2003.

Virtual Archeology at Aphrodisias, a virtual reality installation, was created for Christopher Ratte and his team of NYU archeologists working on the excavation of the ancient city of Aphrodisias—one of the most important archaeological sites of the Greek and Roman periods in Turkey. Using information provided by Ratte's team, Jean-Marc Gauthier, Haluk Goksel, and Zach Rosen created interactive reconstructions of the agora including architectural elements, sculptures, and self-determined virtual characters living inside the virtual city of Aphrodisias. The Aphrodisias project is an exact reconstitution of a day on the agora.

The project is entirely generated by several artificial life systems influencing virtual characters during a 24-hour solar cycle. This experience lasts only 4 minutes for the viewers. Viewers can interact with elements of a Greek market recreated inside the virtual world. There are virtual people, shops, merchants, and animals.

Morning	Afternoon

5 a.m.

6 a.m.

7 a.m.

Twenty-four hours in the life of the agora lasts only 4 minutes for the viewers.

This illustration shows students discussing the location of the stores in the marketplace. Students can view and discuss complex data sets on the multiscreen installation. View of the portable installation built at New York University ITP-Tisch School of the Arts in 2003.

The architectural reconstitution of the Agora.

Figure 5- Activities on the market place include shops, mills with donkeys, and fountains.

The navigation system is based on several interactive cameras. Viewers can activate cameras by moving a "magic wand" in front of the screen. By default, the navigation goes back to the Cocktail camera, a self determined camera. Visitors can also use a P5 3D glove or other infrared device to navigate through the marketplace.

Elements of a Greek market recreated inside the virtual world.

Viewers have two ways they can navigate through the agora. One is by holding a wireless "magic wand"—a wooden stick with reflective paint—in front of the screens. The other is to choose not to interact with the scene and to let the camera become an autonomous Cocktail camera that chooses to follow virtual characters according to their actions and their proximity to the camera. The illustration shows a fully autonomous Cocktail camera filming the interaction between two virtual actors.

Cocktail camera filming the interaction between two virtual actors.

Another significant innovation developed for this project is the Fish-Cyclop camera or multiscreen camera. The Fish-Cyclop camera is an extra behavior that can be added to existing virtual cameras inside virtual environments. The camera can become a stereo camera or a three-camera node at the touch of a key.

Sequence of frames of the agora seen by the Fish-Cyclop camera. The scene is filmed by a node of three virtual cameras attached together.

The low-cost display is made of three screens with a seamless image projected by three video projectors in the space of the installation.

Viewers are invited to stand in front of the three-screen display. Viewers can interact with the virtual world by holding an infrared sensor shaped like a magic wand.

The Fish-Cyclop camera can achieve seamless rendering between several flat-screen panels on a desktop.

5   CONCLUSION

This chapter presents several projects designed around the viewer's experience, with examples of dynamic designs of virtual cameras and sound input or magic wands to control the exploration in a virtual space. The ultimate viewer's experience takes place inside a physical space—the space of the installation or inside a room. The next chapter looks at the physical space of the installation using input devices acting in between the viewer and the projection of the viewer inside a virtual space.

CHAPTER 8

The Viewer's Experience, Interactive Installations, and Displays

"I would like to think of my purpose as a search for what is epic in everyday objects and everyday attitudes."

—Richard Hamilton, artist

1   THE VIEWER'S INTERACTION

Digital projects can take years of constant evolution until the desire to make the artwork visible to the public overrides the complexity inherent to the designing of a virtual world. I do not know how to explain the emotion of watching viewers connect with unfinished sketches of virtual spaces pieced together with code originally scribbled in a coffee shop. Fractions of seconds covered by layers of routine of our everyday life suddenly emerge as interactive art, giving others the feeling of sharing a similar experience. What a great joy to see viewers interact with our creative process, revealing invisible paths, voices, songs, and patterns that were once blurry images stored in our minds.

In the previous chapters of this book we explored how some secret harmony can take place between texts, video, gestures, computer-generated images, movies, virtual spaces, drawings, and hand-drawn sketches. Although harmony in 2D art can be associated with rules of composition and associations between sound and visuals, in the context of installations we also look at the convergence of a virtual space and the physical space surrounding a viewer. As a result of the convergence between the physical and virtual spaces, the spatial design of an installation built around viewers must include sensors or input devices plugged into the virtual world.

The input devices described in this chapter have provided successful interactions between viewers and a virtual world. Input devices with the following criteria are best adapted for controlling or influencing virtual worlds:

1 Aside from being reliable, inexpensive, and easy to set up, the input devices are not constrained to use in a specific location. A sensor that allows the viewer to walk around creates an experience very different from that of a moviegoer. Viewers can walk around the space of the installation and get closer or further from the screen.

2 Input devices should have sensors that can work in spaces of various scales ranging from a room in a museum, to a store window, to a public space in the city.

3 When creating devices, my preferences lean toward solutions that can use several devices or sensors interchangeably. I prefer to stay away from devices that are too specific because we often design an installation in one location and build it in another place with different space constraints that may require flexibility in terms of the setup.

2   SERIAL INPUT DEVICES

In the example of the Infinite City, several viewers walk in front of a large screen set up along one wall of a room; they can perceive that sounds and cameras are changing as they move closer to the screen. The viewers' motions and gestures are captured by input devices. For example, ultrasound sensors detecting the presence of viewers can move the virtual cameras inside the virtual world. Viewers are placed at the center of the dynamic experience of the Infinite City; they are at the intersection of dynamic physical and virtual spaces made of sensors, a room, and the projection of a virtual space on the screen. The viewer communicating with the virtual space starts making decisions about how to navigate though the Infinite City and how to control the motion of the camera. The files from this section can be found on the companion CD-ROM in the Viewer's Experience folder under the Infinite City subfolder.

The illustrations on page 342 show spatial and technical aspects of the installation in a room. The steps along the technologic path show how a sensor capturing the motion of a gesture made by a viewer can process data sent to the virtual world. The same movement is then applied to the virtual camera active in the virtual world. The bottom illustration shows, from left to right, the physical devices and sensors (Steps 1 and 2), the virtual world (Steps 4 through 6), and the 3D acceleration card used for the multiscreen display (Step 7).

2.1 Technological Path for Serial Input Devices for a Virtual World

This chapter presents in-depth discussions about technological paths made of physical devices such as sensors, virtual spaces, and displays. We will discuss how each element of the path plays a specific role inside the viewer's experience. A viewer's experience can be defined with the following points:

- Someone standing or seated in front of a full room-sized screen

- The viewer has a physical presence inside the space of the installation. His or her presence includes motions and gestures captured by input devices, motion sensors, microphones, light sensors, and motion tracking.

- The viewer communicates with the virtual space, making decisions about how to navigate though the content and how to influence the delivery of content.

We will now look at ways to use serial input devices to control a virtual world. Adding a serial input device to Virtools can be done in several ways. One possibility is to write a specific file (in .dll format) that can pass information from one input device to another. See the example of the data glove also covered in this chapter.

The plugin route offers the advantage of making device-specific data handlers that can rely on the manufacturer's SDK for whatever device you are using.

Using Virtools .dll files can have several limitations:

Writing a .dll requires knowledge of the kind of data sent by the input device being used. If you change the input device, you may have to write another .dll.

Frequent software updates in Virtools require the .dlls to be recompiled. This can be costly and can create problems when updating an installation.

There is another solution that may be more appealing: using a Virtools file embedded in Director. This offers more flexibility with serial input devices.

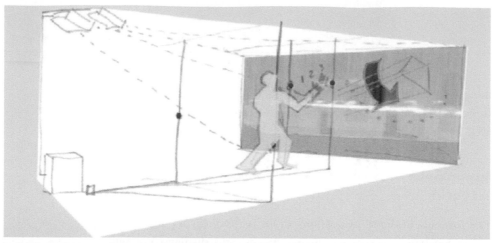

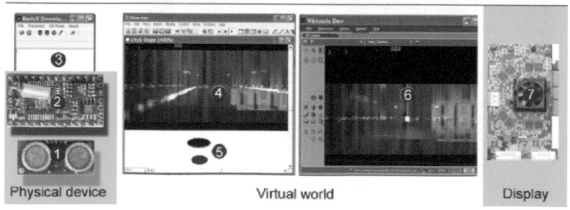

Physical device   Virtual world   Display

Top view: A preliminary sketch of the spatial interaction with the virtual space. Middle view: A viewer interacts with the Infinite City (by Jean-Marc Gauthier, Miro Kirov, and James Tunick). The virtual space is projected in a room. Bottom view: A breakdown of the technological path followed when a sensor captures the motion of a gesture made by a viewer. From left to right: The physical devices and sensors (Steps 1 and 2), the virtual world (Steps 4 through 6), and the 3D acceleration card used for the multiscreen display (Step 7).

The BX microprocessor is a very flexible hardware device that is compatible with many kinds of sensors. Director can process standard data coming from the BX microprocessor and pass the information to Virtools. The same Director Xtras can be reused through software upgrades. The initial setup may look complex, but it pays off in the long run because it can be reused later for testing other kinds of input devices. The following path shows how to connect Virtools, Macromedia Director, the BX microprocessor, and serial input sensors.

2.2 An Example of Virtual Space Installation Using an Ultrasound Sensor

In the following example, we will be setting up an installation that can take place indoors. The short-range ultrasound sensor coupled with a BX microprocessor can work approximately 10 feet from the viewer.

The string of tasks involved with the creative process begins with the design of a virtual world and ends with the placement of sensors in the physical space of the installation. The production workflow is created and tested from the core of the project, the virtual world, then extends to the peripherals, the input devices set up in the physical space of the installation. This creative process is the opposite of the way the installation is finally set up in a room where input devices and peripherals control the virtual world. We will follow the production workflow to understand the technological path that links the virtual world and the physical world.

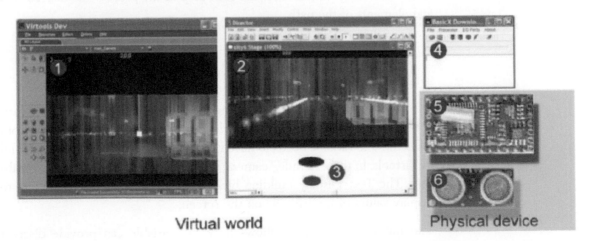

Virtual world   Physical device

The production workflow is created and tested from the core of the project, the virtual world (Steps 1 and 2), then extends to the peripherals, the input devices set up in the physical space of the installation (Steps 4, 5, and 6).

Before having a technological path, we have a collection of hardware and software communicating together. The path is the result of a process in which the main rule is "each new piece is added to the previous ones with the condition that the previous elements of the path are already working well together." In the event of a system breakdown, troubleshooting the last module added to the path should help put the whole system back to work.

Let's walk through the previous illustration:

Step 1. The virtual world called Infinite City is created in Virtools.

Step 2. The Virtools file is embedded inside Director, using a special active control called Virtools Web Player Class.

Step 3. A couple of sliders created in Director's Stage are used to test the functionality between Director and Virtools.

Step 4. The BasicX Express software allows communication between the BX microprocessor and Director. Please note that Director's serial Xtra, made by Geoff Smith, is available at the Physical Bits Web site (http://www.physicalbits.com).

Step 5. This shows the BX microcontroller.

Step 6. The ultrasonic sensor is placed in the same space as the screen where the installation is projected.

In the following section we detail each step of the process and discuss critical elements of the path.

2.3   Passing Variables from the Outside World Inside Virtools

In Virtools, Has Attribute is a building block that can receive values of variables from outside the virtual world—for example, variables sent from Director. The variable needs to have the same name in Director's script and in Virtools. In this example, "camera_move" belongs to a category of attributes. The Text Display building block is used in the following setup to display values of the variable on the screen.

The following illustration shows how the variable can provide discreet values to gradually change the position of the camera or of a 3D object.

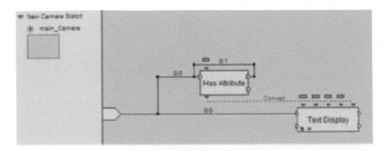

In Virtools, Has Attribute is a building block that can receive values of variables from outside the virtual world—for example, variables sent from Director.

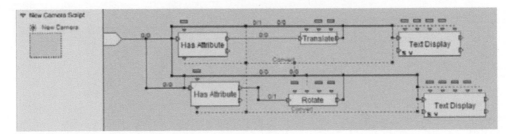

Adding Translate and Rotate building blocks allows the ultrasound sensor to modify the position of the virtual camera inside the virtual world.

In this case, adding Translate and Rotate building blocks allows the ultrasound sensor to modify the position of the virtual camera inside the virtual world.

The following illustration shows how three different sensors can pass values inside a virtual world. Each variable activates a different modifier—in this case, movements for cameras and particle animation.

A sensor can also trigger a message to switch cameras—for example, switching the active camera in Virtools with a smooth transition between the yellow path camera to the blue path camera. The values of the variable common to Director and Virtools are constantly tested against a threshold value. When the values of the variable are above the threshold, the behavior will trigger an action. The value of the threshold is determined by the Test building block. The following example shows how to use a threshold to send a message with the Broadcast Message building block. In this case the message is to switch path cameras and to use a transition camera for a smooth transition between the yellow path camera to the blue path camera.

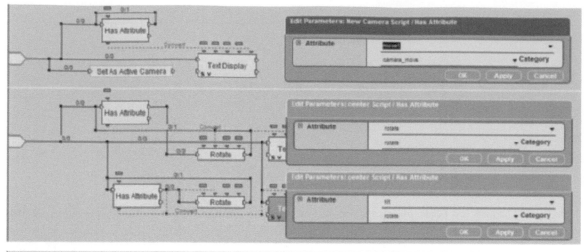

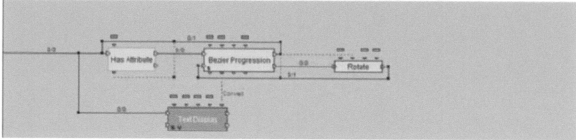

Top view: Three different sensors pass values inside a virtual world. Each variable activates a different Has Attribute building block and a different modifier—in this case, movements for cameras and particle animation. Bottom view: This setup adds a delay to camera movements. This allows the viewer to see the camera reacting to a user input. The camera behavior shown was designed for the installation Infinite City.

For values under the threshold, the camera remains unchanged. For values above the threshold, the active camera switches to the camera on the blue path.

Setting up a threshold in Virtools requires several steps.

Step 1. The Has Attribute building block reads a variable from the outside world as it is passed into the virtual world.

Step 2. The value is tested against the threshold. If the test is positive,

Step 3. A message is sent on to the virtual camera behavior.

Step 4. The message is received inside another script, designed for the transition camera. The message is accepted by the Wait Message building block that triggers the transition camera setup. Please note that, in this example, before hooking up the sensors we use a slider in Director to test communication between Virtools and Director.

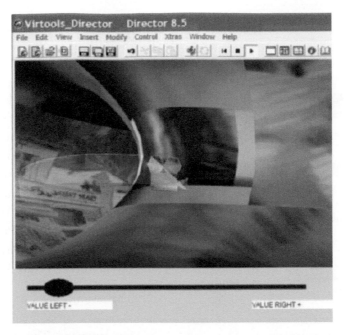

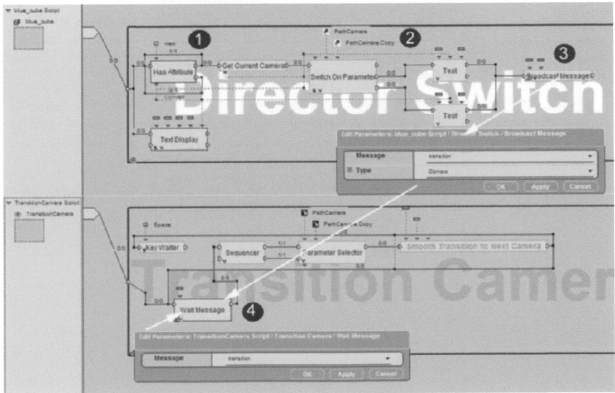

Top view: Moving the slider in Director can switch the directions of cameras in Virtools. The bottom view shows how Virtools' behaviors react to the changes of value for the variable coming from Director. Step 1: A variable from the outside world is passed inside the virtual world. Step 2: The value is tested against the threshold. If the test is positive, Step 3, a message is sent on. Step 4: The message is received inside another script, designed for the transition camera.

2.4   Embedding a Virtools Scene Inside a Director Movie

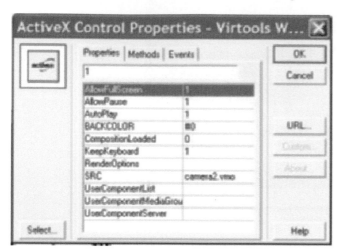

The Active X Control Properties window lets you define many playback parameters and the file name of the Virtools content.

Embedding a Virtools scene inside Director requires two levels of setup. The first level of setup ensures that the Director and Virtools worlds can communicate using Active X. The Active X Control Properties can be found in Director under Insert > Control > ActiveX > Virtools Web Player Class. The Active X Control Properties window lets you define many playback parameters and the file name of the Virtools content.

The second level of setup requires the creation of a variable in Director's Lingo scripting language. This variable is also used in Virtools' Has Attribute building block. The following illustration shows how the variable sprLocH is tested with sliders in Director's stage window. Values of the variable are displayed with no latency inside the Virtools window, confirming that the setup is working.

Once you have tested your setup with one or two variables it can be implemented with many variables. For example, the following are five variables used to control the camera and a particle generator in the following Lingo code:

```
global rotateV,tiltV,transV,magnit,vortex,cameraV,var1,var2,var3,var4

on exitFrame me

  myfunc

  sprite(1).DoCommand("SetAttribute" & " 'center' " & "'rotate'"&
rotateV )

  sprite(1).DoCommand("SetAttribute" & " 'main_Camera' " &
"'camera'"& cameraV )

  sprite(1).DoCommand("SetAttribute" & " 'center' " & "'rotateY'"&
transV )

  sprite(1).DoCommand("SetAttribute" & " 'mangnit' " & "'Particle
Magnet'"& magnit )
```

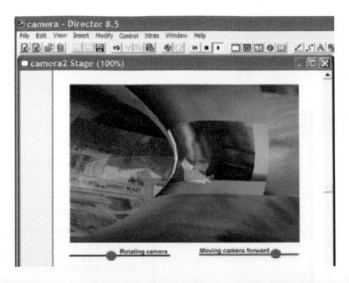

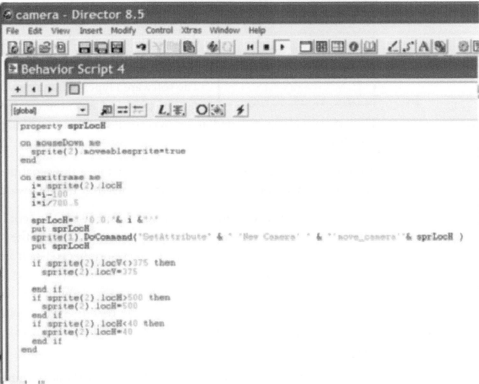

Top view: The variable sprLocH is tested with sliders in Director's stage window. Bottom view: Values of the variable are displayed with no latency inside the Virtools window, confirming that the setup is working.

Interface for the BasicX software.

sprite(1).DoCommand("SetAttribute" & " 'vortex2' " & "'Particle Vortex'"& vortex)

go the frame

end

2.5   Serial Communication

Setting up a serial communication protocol on your computer requires several steps. The Hyperterminal software—or BasicX if you use a BX microprocessor—helps to establish serial communication. These software packages can open a communication port, called Com port, and monitor the flow of serial data coming from the input device—for example, the BX microprocessor.

The software Hyperterminal lets you choose a Com port and monitor the flow of data. Hyperterminal can be found under Accessories > Communication > Hyperterminal.

BasicX is a software package that lets you choose the serial port and lets you close other Com ports. Com port 2 is used in the following example.

Director receives data coming to the serial port using the serial extra that can be found at www.physicalbits.com. Remember to add the serial extra to Director's Xtra folder. In Director, a Lingo script gets data from the serial port. In this example, the line of code gSerialPort = (new(xtra "SerialXtra," "com2") specifies Com port 2 as the serial port. The line of code var1=gSerialPort.readNumber( ) assigns the data to specific variables. The following Lingo code will make sure to retrieve data coming from the serial port:

```
global gSerialPort,var1,var2,var3, var4,var,rotateV,transV,cameraV,
vortex,magnit

on startMovie

clearglobals
```

```
  openxlib "SerialXtra"

  gSerialPort = new( xtra "SerialXtra," "com2" )    --specify which port.

  gSerialPort.SetProtocol(9600, "n,"8,1)

  gSerialPort.readString( )

  gSerialPort.writeString("Z")

  flushInputBuffer(gSerialPort)

end

on stopMovie

  gSerialPort = 0

end

on myfunc

  if objectP(gSerialPort)   then

    repeat while (charsAvailable(gSerialPort) >1)

      var= gSerialPort.readNumber( )

      if var=0 then

        var1=gSerialPort.readNumber( )

        var1=(var1/10)-25

        var1=var1/-1000.0

      else if var =1 then

        var2=gSerialPort.readNumber( )

        var2=(var2/10)-25

        var2=var2/-1000.0

      else if var=2 then

        var3=gSerialPort.readNumber( )

        var3=(var3/10)-25

        var3=var3/-1000.0
```

```
        else if var=3 then
            var4=gSerialPort.readNumber( )
            var4=(var4/10)-25
            vortex=var4/1000.0
            magnit=var4/100.0
        end if

        rotateV=" '"& var1 &"'"
        transV=" '"&var2&"'"
        cameraV=" '"& var3 & "'"
        magnit=" '"& var4 & "'"
        vortex= " '"& var4 &,'" & var4 & "'"

        if var1 or var2 or var3 or var4<0.001 then
            sprite(1).DoCommand("SetAttribute" & " 'center' " & "'IC'"&
"'center'" )
            end if

    end repeat
    end if
    end
```

2.6   Sensors

The sensor and the BX microprocessor are the last elements of the technological path. Several ultrasonic sensors placed in the space of the installation give a multidimensional presence. The BX microcontroller stands between the sensors and the connection to the computer's serial port. The BX microcontroller requires the downloading of the following software in the BX chip and a 12-volt power source.

```
Option Explicit
' setup input pin
'for the ultrasonic ranger
```

```
const TrigPin As Byte = 13
const TrigPin2 As Byte = 15
const TrigPin3 As Byte = 17
const TrigPin4 As Byte = 19

const EchoPin As Byte = 14
const EchoPin2 As Byte = 16
const EchoPin3 As Byte = 18
const EchoPin4 As Byte = 20

dim a1 As Byte
dim a2 As Byte
dim a3 As Byte
dim a4 As Byte

dim Range as Integer
dim Range2 as Integer
dim Range3 as Integer
dim Range4 as Integer

dim ranger as integer
dim ranger2 as integer
dim ranger3 as integer
dim ranger4 as integer

dim invar as byte

Public Sub Main( )

 ' red light
 putpin 25, 0
 putpin 26, 1
 delay 0.5
```

```
' initial variable states:

dim outputBuffer(1 to 10) as byte
dim inputBuffer(1 to 10) as byte

 call defineCom3(12, 11, bx1000_1000)
 call openQueue(inputBuffer, 10)
 call openQueue(outputBuffer, 10)
 call openCom(3, 9600, inputBuffer, outputBuffer)

' green light
 putpin 25, 1
 putpin 26, 0

'ultrasonic ranger
call PutPin(EchoPin, bxInputTristate)
call PutPin(TrigPin, bxOutputLow)

'ultrasonic ranger2
call PutPin(EchoPin2, bxInputTristate)
call PutPin(TrigPin2, bxOutputLow)

'ultrasonic ranger3
call PutPin(EchoPin3, bxInputTristate)
call PutPin(TrigPin3, bxOutputLow)

'ultrasonic ranger4
call PutPin(EchoPin4, bxInputTristate)
call PutPin(TrigPin4, bxOutputLow)

do
a1 = 0
a2 = 1
a3 = 2
a4 = 3
```

```
'ultrasonic ranger

call PulseOut(TrigPin, 10, 1)

Range = PulseIn(EchoPin, 1) \ 54


ranger = getADC(13)

ranger = cInt(cSng(Range) * 2.0)

if ranger < 4 then

ranger = 4

end if

if ranger > 255 then

ranger = 255

end if

call putQueue(OutputBuffer,a1,1)

call putQueue(OutputBuffer,ranger,1)


'debug.print "a1" ; cstr(a1)

'debug.print "ranger1" ; cstr(ranger)


'ultrasonic ranger 2

call PulseOut(TrigPin2, 10, 1)

Range2 = PulseIn(EchoPin2, 1) \ 54


ranger2 = getADC(15)

ranger2 = cInt(cSng(Range2) * 2.0)

if ranger2 < 4 then

ranger2 = 4

end if

if ranger2 > 255 then

ranger2 = 255
```

```
end if

call putQueue(OutputBuffer,a2,1)

call putQueue(OutputBuffer,ranger2,1)

'debug.print "a2" ; cstr(a2)

' debug.print "ranger2" ; cstr(ranger2)

'ultrasonic ranger 3

call PulseOut(TrigPin3, 10, 1)

Range3 = PulseIn(EchoPin3, 1) \ 54

ranger3 = getADC(17)

ranger3 = cInt(cSng(Range3) * 2.0)

if ranger3 < 4 then

ranger3 = 4

end if

if ranger3 > 255 then

ranger3 = 255

end if

call putQueue(OutputBuffer,a3,1)

call putQueue(OutputBuffer,ranger3,1)

'debug.print "a3" ; cstr(a3)

'debug.print "ranger3" ; cstr(ranger3)

'ultrasonic ranger 4

call PulseOut(TrigPin4, 10, 1)

Range4 = PulseIn(EchoPin4, 1) \ 54

ranger4 = getADC(19)

ranger4 = cInt(cSng(Range4) * 2.0)
```

```
if ranger4 < 4 then

ranger4 = 4

end if

if ranger4 > 255 then

ranger4 = 255

end if

call putQueue(OutputBuffer,a4,1)

call putQueue(OutputBuffer,ranger4,1)

debug.print "a4" ; cstr(a4)

debug.print "ranger4" ; cstr(ranger4)

call delay(0.02)

loop

end sub
```

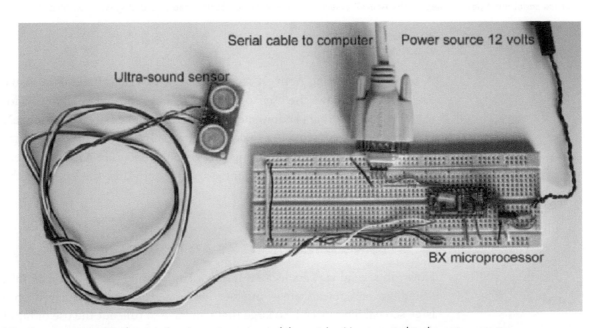

BX microprocessor setup showing the ultrasonic sensor and the serial cable connected to the main computer.

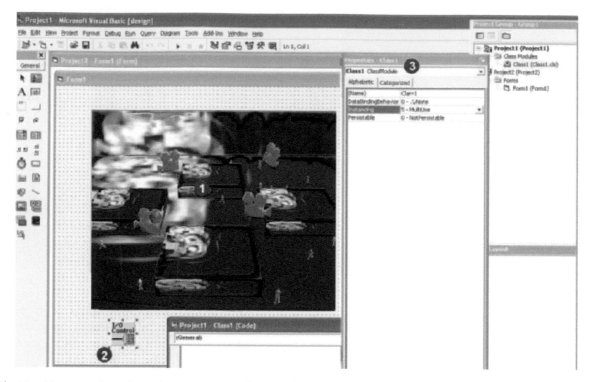

The Visual Basic interface allows the user to create drag-and-drop ActiveX objects for the Virtools rendering window, Step 1, and for the serial input device, Step 2. The ActiveX object I/O control can be customized in a dialog window, Step 3.

2.7   Using Active X with Microsoft Visual Basic®

A similar setup using ActiveX can be created in Visual Basic. The Visual Basic interface allows the user to create drag-and-drop ActiveX objects for the Virtools rendering window and for the serial input device. Each ActiveX object can be customized in a dialog window. The resulting project is compiled into an executable file. Using the Virtools rendering AX component takes advantage of VB's Serial controls, which obviates the need for the Director instance.

3   USING A TELEPHONE AS AN INPUT DEVICE

How can several viewers located inside a large public space—for example, a stadium or sidewalk area—interact with a virtual space displayed on a large screen? The space is so large that you have to rule out the use of input devices such as a mouse or keyboard or even devices like the ultrasound sensors. A cellular phone may be the answer. People already carry the device with them, they know how to use it, and they only need to dial a number to get connected with the virtual world. Cellular phones can be used while the viewer is walking in the proximity of the screen. The cellular phone allows sound feedback from inside the virtual world. The interactive demo can be found on the companion CD-ROM in the Viewer's Experience folder under Nighthawks.cmo.

The idea of working with telephones, ground lines, or cellular phones was inspired by watching an interactive television experiment. Using a telephone seemed to be an easy way for a television viewer to interact with on-air content while watching TV. The device used for interactive television is a DTMF (dual-tone multifrequency) converter, which takes an analog signal from the phone and converts it into serial communication. This converter can also work two ways when a digital sound coming from a virtual world can be broadcasted over the phone line. The DTMF converter is connected to a ground phone line through an answering machine that can receive calls and play a message explaining to people that they are about to interact with a virtual world. At the end of the message, the incoming call opens a channel inside the virtual world and allows the caller to use the phone keypad as a remote control. Proximity of the elements of the setup is a key issue when using a phone setup. The DTMF converter needs to be close to the computer generating the virtual world. The computer will also need to be near the screen display or display element of choice.

3.1   Technological Path

The technological path for setting up a phone as an input device in Virtools requires several steps.

Step 1. A ground-line phone or cellular phone is used to call the converter through an answering machine.

Step 2. The DTMF converter produced by SIA Sistemas is designed to allow up to four simultaneous phone lines.

Step 3. This step shows the serial port connecting the converter to the computer.

Step 4. Hyperterminal software is used to open a Com port and to monitor the serial communication.

Physical device   Virtual world   Display

Setting up a phone as an input device requires several steps. In Step 1, a ground-line phone or cellular phone is used to call the converter through an answering machine. In Step 2, the DTMF converter designed by SIA Sistemas allows up to four simultaneous phone lines. Step 3 shows the serial port connecting the converter to the computer. In Step 4, Hyperterminal software is used to open a Com port and to monitor the serial communication. Step 5 shows Virtools receiving the serial data through the SIA TV building block provided by SIA Sistemas. In Step 6, the 3D acceleration card renders the virtual world presented on the display.

Step 5. This step shows Virtools receiving the serial data through the SIA TV building block provided by SIA Sistemas.

Step 6. The 3D acceleration card renders the virtual world presented on the display.

3.2   An Example of a Virtual Space Installation Using a Telephone in Virtools

The script illustrated on page 362 shows how to control an interactive character during a telephone call. The script includes a string of keyboard controls behaviors for easy testing of the character and a string of SIA TV behaviors for remote control by a telephone.

Creating multiple controls is very useful for an installation. In the following example, a character can be controlled three ways: by a keyboard, by a

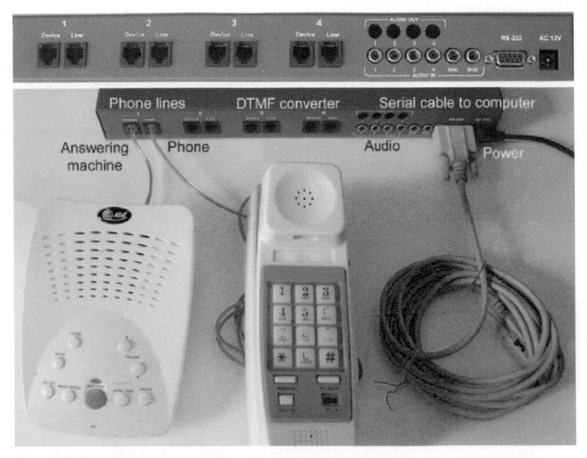

DTMF converter setup showing an answering machine, a phone receiver, and the serial cable connected to the main computer.

phone, or by an artificial intelligence system. In this instance, the keyboard system is used for troubleshooting, the phone acts as the viewer's input device, and the artificial intelligence system takes over the virtual world and directs characters when no input is sent from the viewer. If the phone line is not available, the keyboard replaces its role as the input device. If there is no input at all from a viewer, the artificial intelligence system will take control. This redundancy of elements creates the feeling of an ongoing virtual world that can exist without viewer input. Viewers discovering the installation feel that they are naturally invited to participate in the life of the virtual world that evolves in front of them.

Let's look at the setup of the building blocks just described. The parameters of the SIA TV behavior include the following:

- Choosing a Com port for the serial input and the phone line being used. Please note that this SIA TV behavior can assign a Com port and bypass the Hyperterminal software.

- The various output pins of the Wait Data building block correspond to the keys of a phone keypad.

- Selected output pins are used to broadcast a message to the Unlimited Controller building block located inside the character's script.

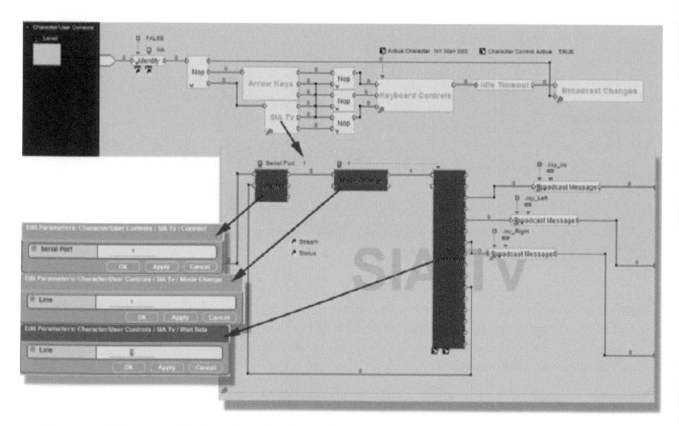

This script includes a string of keyboard controls behaviors for easy testing of the character and a string of SIA TV building blocks for remote control by a telephone. The parameters of the SIA TV behavior include choosing a Com port for the serial input and the phone line being used. The various output pins of the Wait Data building block correspond to the keys of a phone keypad. Selected output pins are used to broadcast a message to the Unlimited Controller building block located inside the character's script.

4   MUSICAL INSTRUMENT DIGITAL INTERFACE

Musical Instrument Digital Interface (MIDI) is a communication protocol that gives a digital representation of a sound including values for the note's pitch, length, volume, attack, and delay time. MIDI can be used to create music and to exchange data between a MIDI keyboard and a computer or between MIDI-compatible software. MIDI can be used to send real-time data between a software processing sound and video—for example, MAX-msp or jitter—to a virtual world. MIDI provides an ideal interface for creating real-time associations between music, video tracking, images, and virtual spaces.

4.1   Setting Up MIDI Channels

The first step is to set up a MIDI channel for data flowing from the MIDI software into the virtual world. Channels are setup paths for communication on a computer. Generally, computers use only one MIDI channel dedicated to the sound card. The most common channels for MIDI are 0, 1, or 2. To access MIDI channels on your computer and connect Max-MSP and Virtools, you will need to use MIDI Yoke software (available at www.midiox.com). Please note that Max will use MIDI Yoke 1 as an output port and Virtools will use channel 1 as an input port.

4.2   Sensor and Microcontroller

The second step is to create the physical setup. The input device can be a microcontroller with sensors connected to the computer. In the following illustration we use a Teleo microcontroller (from www.makingthings.com) with an infrared sensor. The microcontroller is connected to the computer with a USB connection.

4.3   Virtools and Max Configuration

The second step is to set up a MIDI input port in Virtools to allow communication using the MIDI channel. In the example on page 365 the software Max-MSP sends MIDI data with two parameters, pitch and volume, that can be received in Virtools. The Set MIDI Input Port building block checks the MIDI input port, and Read MIDI checks the data. This building block may help you determine which channel and note MIDI is using. The Read MIDI building block is activated once a note gets in.

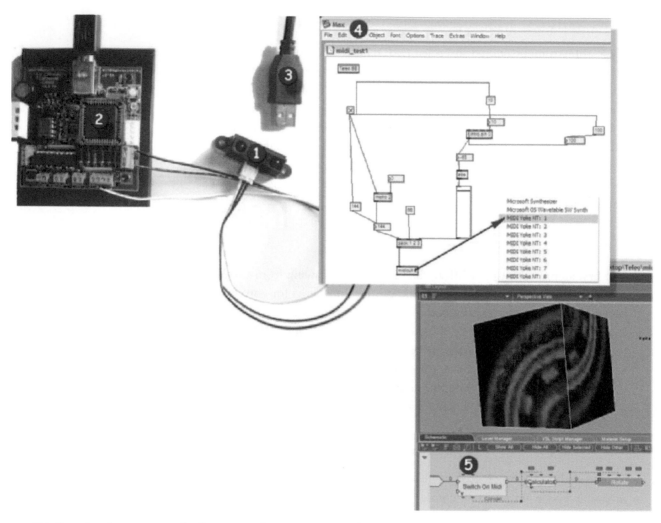

This illustration shows an example of connection between sensors and Virtools using MIDI. The chain of hardware and software includes an infrared sensor (1), a microcontroller (2), a USB connection (3), the software Max-MSP (4), and Virtools (5). The microcontroller shown here is the Teleo (from www.makingthings.com).

The actions controlled by the flow of MIDI data can be dispatched to the Switch on Parameter building block that will trigger various strings of behaviors. The Switch on MIDI building block is set up with the MIDI parameters channel. This reveals which channel is parsed by the building block. Please check the step-by-step tutorial on sensors using MIDI, which can be found on the companion CD-ROM.

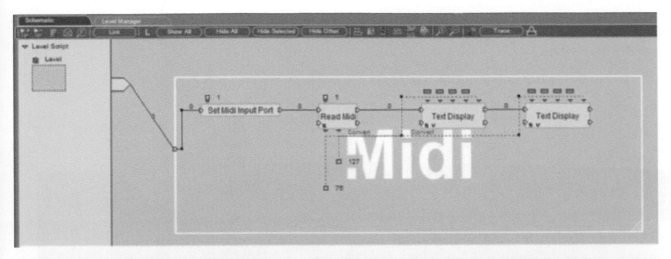

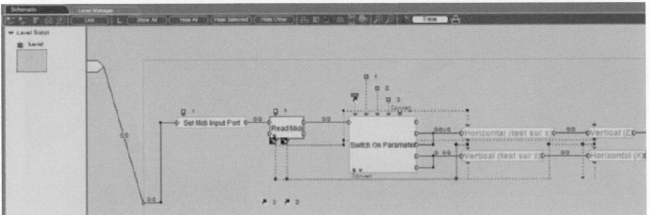

Max-MSP sends MIDI data with two parameters, pitch and volume, that can be received in Virtools. The Set MIDI Input Port building block checks the MIDI input port, and Read MIDI checks the data. The actions controlled by the flow of MIDI data can be dispatched to the Switch on Parameter building block that will trigger various strings of behaviors.

5   DATA GLOVE

A data glove can process gestures from the hand and turn them into data for navigation inside a virtual world. This section shows how a data glove is used as an input device to control the motion of a virtual hand. We add physics to the virtual hand, so it can actually slap objects like balls. The interactive demo can be found on the companion CD-ROM in the Viewer's Experience folder under the Hand subfolder.

The data glove is a low-cost input device combining flex-sensors on each finger and infrared sensors in the palm. Although the glove was designed

A data glove is used as an input device to control the motion of a virtual hand. We add physics to the virtual hand, so it can actually slap objects like balls. (Project created by Mike Olson.)

to guide the mouse pointer on the screen and to override the mouse movements, we use a version of the glove that sends data directly to the virtual world. The P5 glove, manufactured by Essential Reality, is the one referred to in this section. The glove is connected to a computer via a USB port. Flex-sensors mounted on each finger track how far the fingers bend. Infrared sensors mounted around the palm of the glove follow the motion of the hand on a 2D plane. We use the P5 glove with Mike Olson's GetP5Data building block to read data coming from the glove.

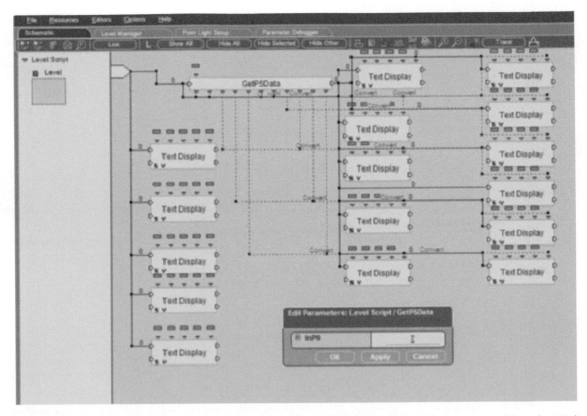

Testing the streams of data coming from the glove. The Text Display building block displays the data received from the P5 glove inside the 3D Layout window.

This is how it works. The first step is to test the streams of data coming from the glove. The Text Display building block displays data received from the P5 glove in the 3D Layout window.

The second step is to create a virtual hand with bones and joints. Please refer to the Chapter 4 for more details about applying inverse kinematics. The virtual hand is rigged to the P5 glove.

Note: The following text is a sample of code for the data glove. The full version of the code can be found in the folder for Chapter 8 on the companion CD-ROM.

```
//////////////////////////////////////////////////

//              GetP5Data

//////////////////////////////////////////////////
```

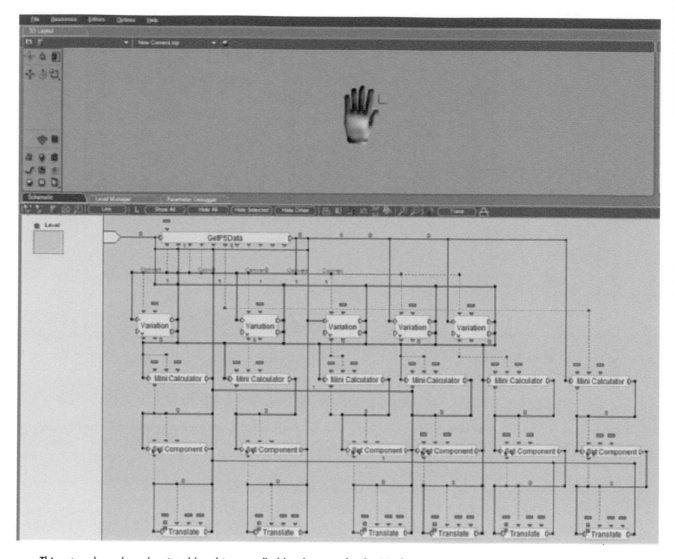

This setup shows how the virtual hand is controlled by data sent by the P5 glove.

```
include "CKAll.h"
include "P5dll.h"

CKObjectDeclaration *FillBehaviorGetP5DataDecl( );
CKERROR CreateGetP5DataProto(CKBehaviorPrototype **);
int GetP5Data(const CKBehaviorContext& BehContext);
```

```
//p5 creation
CP5DLL P5;

//These variables contain the actual X, Y, Z position of the cursor.
int nXPos = 0, nYPos = 0, nZPos = 0;

//These variables contain the frame to frame deltas of the cursor.
float fXMickey = 0.0f, fYMickey = 0.0f, fZMickey = 0.0f;

//These variables contain the filtered orientation information.
float fAbsYawPos, fAbsPitchPos, fAbsRollPos;
float fRelYawPos, fRelPitchPos, fRelRollPos;

float fXPos[P5MOTION_XYZFILTERSIZE, fYPos[P5MOTION_
XYZFILTERSIZE, fZPos[P5MOTION_XYZFILTERSIZE;
float fFilterX, fFilterY, fFilterZ;

CKObjectDeclaration *FillBehaviorGetP5DataDecl( )
{
   CKObjectDeclaration *od =
CreateCKObjectDeclaration("GetP5Data");

   od->SetType(CKDLL_BEHAVIORPROTOTYPE);
   od->SetVersion(0x00010000);
   od->SetCreationFunction(CreateGetP5DataProto);
   od->SetDescription("Gets Data From P5 Glove");
   od->SetCategory("UserBBs");
   od->SetGuid(CKGUID(0xD515125D,0x402CA9A8));
   od->SetAuthorGuid(CKGUID(0x56495254,0x4f4f4c53));
   od->SetAuthorName("Mike Olson");
   od->SetCompatibleClassId(CKCID_BEOBJECT);

   return od;
}
```

```
CKERROR CreateGetP5DataProto(CKBehaviorPrototype** pproto)
{
  CKBehaviorPrototype *proto =
CreateCKBehaviorPrototype("GetP5Data");
  if(!proto)      return CKERR_OUTOFMEMORY;

//---    Inputs Declaration
      proto->DeclareInput("In0");

//---    Outputs Declaration
      proto->DeclareOutput("Out0");

//-----  Input Parameters Declaration
      proto->DeclareInParameter("InP0","CKPGUID_INT");

//---    Output Parameters Declaration
      proto->DeclareOutParameter("OutP0","CKPGUID_INT");
      proto->DeclareOutParameter("OutP1","CKPGUID_INT");
      proto->DeclareOutParameter("OutP2","CKPGUID_INT");
      proto->DeclareOutParameter("OutP3","CKPGUID_INT");
      proto->DeclareOutParameter("OutP4","CKPGUID_INT");
      proto->DeclareOutParameter("OutP5","CKPGUID_FLOAT");
      proto->DeclareOutParameter("OutP6","CKPGUID_FLOAT");
      proto->DeclareOutParameter("OutP7","CKPGUID_FLOAT");
      proto->DeclareOutParameter("OutP8","CKPGUID_FLOAT");
      proto->DeclareOutParameter("OutP9","CKPGUID_FLOAT");
      proto->DeclareOutParameter("OutP10","CKPGUID_ FLOAT");

    P5.P5_Init( );
    P5.P5_SetMouseState(0, false);
```

```
    //----   Local Parameters Declaration

    //----   Settings Declaration

      proto->SetFlags(CK_BEHAVIORPROTOTYPE_NORMAL);

      proto->SetFunction(GetP5Data);

    *pproto = proto;

    return CK_OK;

}

int GetP5Data(const CKBehaviorContext& BehContext)

{

    CKBehavior* beh = BehContext.Behavior;

    beh->ActivateInput(0,FALSE);

    int deltat = 0;

    int ret0 = 0;

    int ret1 = 0;

    int ret2 = 0;

    int ret3 = 0;

    int ret4 = 0;

    float retx = 0;

    float rety = 0;

    float retz = 0;

    float retyaw = 0;

    float retpitch = 0;

    float retroll = 0;
```

```
beh->GetInputParameterValue(0,&deltat);

P5Motion_Process( );

ret0=(P5.m_P5Devices[0.m_byBendSensor_Data[P5_THUMB);

ret1=(P5.m_P5Devices[0.m_byBendSensor_Data[P5_INDEX);

ret2=(P5.m_P5Devices[0.m_byBendSensor_Data[P5_MIDDLE);

ret3=(P5.m_P5Devices[0.m_byBendSensor_Data[P5_RING);

ret4=(P5.m_P5Devices[0.m_byBendSensor_Data[P5_PINKY);

retx = fFilterX;
rety = fFilterY;
retz = fFilterZ;

retyaw = fAbsYawPos / float(90.0);
retpitch = fAbsPitchPos / float(90.0);
retroll = fAbsRollPos / float (180.0);

beh->SetOutputParameterValue(0, &ret0);
beh->SetOutputParameterValue(1, &ret1);
beh->SetOutputParameterValue(2, &ret2);
beh->SetOutputParameterValue(3, &ret3);
beh->SetOutputParameterValue(4, &ret4);

beh->SetOutputParameterValue(5, &retx);
beh->SetOutputParameterValue(6, &rety);
beh->SetOutputParameterValue(7, &retz);

beh->SetOutputParameterValue(8, &retyaw);
beh->SetOutputParameterValue(9, &retpitch);
```

```
  beh->SetOutputParameterValue(10, &retroll);

  beh->ActivateOutput(0);

  return CKBR_OK;

}

/***********************************************

Function: P5Motion_FilterXYZ( )

Use: Internal Function.  Used to filter XYZ Data

Parameter: None

***********************************************/

void P5Motion_FilterXYZ( )

{

  static int firsttime = 1;

  //if (P5 != NULL)

  //{

    if (firsttime==1)

    {

      //Don't process anything on the first call, just init our filter arrays

      for (int i=0; i<P5MOTION_XYZFILTERSIZE; i++)

      {

        fXPos[i = P5.m_P5Devices[0.m_fx;

        fYPos[i = P5.m_P5Devices[0.m_fy;

        fZPos[i = P5.m_P5Devices[0.m_fz;

      }

      firsttime = 0;

    }
```

else

{

//We are going to implement a dynamic filter, which will flush the filter array values at different rates based on the rate of change of the user's hand.

//This will allow for greater latency of motion.

//The setpoint determines the number of pixel motion that will flush the entire filter.

//The idea is when the user does not move much, we average a lot of frames, but during fast motion, we average fewer and fewer frames to reduce latency.

```
define FLUSH_SETPOINT   30.0f

float xflushsize, yflushsize, zflushsize;

int i, j;
```

//Let's determine the number of frames we intend to average together.

```
xflushsize = fabs(P5.m_P5Devices[0.m_fx -
fXPos[P5MOTION_XYZFILTERSIZE-1)/2.0f;

    xflushsize *= P5MOTION_XYZFILTERSIZE/FLUSH_
    SETPOINT;

    xflushsize = floor(xflushsize+1.0f);

    if (xflushsize>(P5MOTION_XYZFILTERSIZE-1))

        xflushsize = P5MOTION_XYZFILTERSIZE-1;

yflushsize = fabs(P5.m_P5Devices[0.m_fy - fYPos[P5MOTION_
XYZFILTERSIZE-1)/2.0f;

    yflushsize *= P5MOTION_XYZFILTERSIZE/FLUSH_
    SETPOINT;
```

```
        yflushsize = floor(yflushsize+1.0f);

        if (yflushsize>(P5MOTION_XYZFILTERSIZE-1))

          yflushsize = P5MOTION_XYZFILTERSIZE-1;

  zflushsize = fabs(P5.m_P5Devices[0.m_fz - fZPos[P5MOTION_
XYZFILTERSIZE-1)/2.0f;

        zflushsize *= P5MOTION_XYZFILTERSIZE/FLUSH_
    SETPOINT;

        zflushsize = floor(zflushsize+1.0f);

        if (zflushsize>(P5MOTION_XYZFILTERSIZE-1))

          zflushsize = P5MOTION_XYZFILTERSIZE-1;

//Flush the array based on the number of values determined before.
        for (j=0; j<(int)(xflushsize); j++)

      {

        for (i=0; i<(P5MOTION_XYZFILTERSIZE-1); i++)

        {

          fXPos[i = fXPos[i+1;

        }

      fXPos[P5MOTION_XYZFILTERSIZE-1 =
      P5.m_P5Devices[0.m_fx;

      }

      for (j=0; j<(int)(yflushsize); j++)

      {

        for (i=0; i<(P5MOTION_XYZFILTERSIZE-1); i++)
```

```
            {
                fYPos[i = fYPos[i+1];
            }
        fYPos[P5MOTION_XYZFILTERSIZE-1 =
        P5.m_P5Devices[0.m_fy;
        }
        for (j=0; j<(int)(zflushsize); j++)
        {
            for (i=0; i<(P5MOTION_XYZFILTERSIZE-1); i++)
            {
                fZPos[i = fZPos[i+1];
            }
        fZPos[P5MOTION_XYZFILTERSIZE-1 =
        P5.m_P5Devices[0.m_fz;
        }
    }
    //Average all the values in the filter to smooth the data.
    fFilterX = 0.0f;
    fFilterY = 0.0f;
    fFilterZ = 0.0f;
    for (int i=0; i<P5MOTION_XYZFILTERSIZE; i++)
    {
        //We are going to divide the values to get rid of some jitter.
        fFilterX += fXPos[i/2.0f;
        fFilterY += fYPos[i/2.0f;
        fFilterZ += fZPos[i/2.0f;
    }
```

```
    fFilterX /= P5MOTION_XYZFILTERSIZE;

    fFilterY /= P5MOTION_XYZFILTERSIZE;

    fFilterZ /= P5MOTION_XYZFILTERSIZE;

//  }

}



/***********************************************

Function: P5Motion_FilterYPR( )

Use: Internal Function.  Used to filter Orientation Data

Parameter: None

***********************************************/


float fYaw[P5MOTION_YPRFILTERSIZE, fPitch[P5MOTION_
YPRFILTERSIZE,fRoll[P5MOTION_YPRFILTERSIZE;

float fFilterYaw, fFilterPitch, fFilterRoll;

void P5Motion_FilterYPR( )

{
    static int firsttime = 1;

        if (firsttime==1)

        {

            //Don't process anything on the first call, just init our filter arrays.

            for (int i=0; i<P5MOTION_YPRFILTERSIZE; i++)

            {
```

```
            fYaw[i = P5.m_P5Devices[0.m_fyaw;

            fPitch[i = P5.m_P5Devices[0.m_fpitch;

            fRoll[i = P5.m_P5Devices[0.m_froll;

        }

      firsttime = 0;

   }

   else

   {

      for (int i=0; i<(P5MOTION_YPRFILTERSIZE-1); i++)

      {

         fYaw[i = fYaw[i+1;

         fPitch[i = fPitch[i+1;

         fRoll[i = fRoll[i+1;

      }

      if (P5.m_P5Devices[0.m_fyaw < 91 && P5.m_P5Devices[0.m_
      fyaw> -91)

         {

      fYaw[P5MOTION_YPRFILTERSIZE-1 =
      P5.m_P5Devices[0.m_fyaw;

         }

      if (P5.m_P5Devices[0.m_fpitch < 91 && P5.m_P5Devices[0.m_
      fpitch> -91)

         {

      fPitch[P5MOTION_YPRFILTERSIZE-1 =
      P5.m_P5Devices[0.m_fpitch;

         }
```

```
    fRoll[P5MOTION_YPRFILTERSIZE-1 =
P5.m_P5Devices[0.m_froll;

    }

//Average all the values in the filter to smooth the data.

fFilterYaw = 0.0f;

fFilterPitch = 0.0f;

fFilterRoll = 0.0f;

for (int i=0; i<P5MOTION_YPRFILTERSIZE; i++)

{

    fFilterYaw += fYaw[i;

    fFilterPitch += fPitch[i;

    fFilterRoll += fRoll[i;

}

fFilterYaw /= P5MOTION_YPRFILTERSIZE;

fFilterPitch /= P5MOTION_YPRFILTERSIZE;

fFilterRoll /= P5MOTION_YPRFILTERSIZE;

//  }

}


/*********************************************

Function: P5Motion_Process( )

Use: Processes the XYZ motion information every frame. Call this func-
tion  to calculate a pointer's position and orientation with filtering and
acceleration.

Parameter: None

*********************************************/
```

```
void P5Motion_Process( )

{

  if (P5.m_P5Devices[0.m_byButtons[0 ||
  P5.m_P5Devices[0.m_byButtons[1]

   {

     P5.P5_Init( );

     //flush filters

     for (int i=0; i<P5MOTION_YPRFILTERSIZE; i++)

        {

          fYaw[i = 0;

          fPitch[i = 0;

          fRoll[i = 0;

        }

     for ( i=0; i<P5MOTION_XYZFILTERSIZE; i++)

        {

          fXPos[i = 0;

          fYPos[i = 0;

          fZPos[i = 0;

        }

     fAbsYawPos = 0;

     fAbsPitchPos = 0;

     fAbsRollPos = 0;
```

```
        fFilterX=0;

        fFilterY=0;

        fFilterZ=0;

    }

}
```

6   INFRARED SENSORS AND MOUSE-DRIVEN DEVICES

People should be able to walk freely though the space of an installation without the need to be standing next to a mouse pad. However, sometimes you will need a viewer to use a mouse for the installation. An infrared device is a natural wireless extension for the mouse controls. The infrared sensor made by the company Natural Point follows the motion of a reflective material held by the viewer. The reflective surface placed on a stick, a baseball cap, or even a ring can take control of the motion of the mouse pointer on the screen. It also registers mouse clicks when combined with a foot pedal. The device has a USB connector. Please refer to the section on mouse navigation for more information about how to script mouse inputs in Virtools.

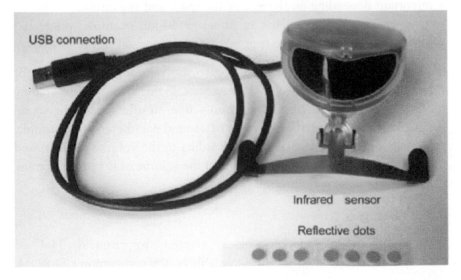

The infrared sensor made by the company Natural Point follows the motion of a reflective material held by the viewer. The reflective surface placed on a stick, a baseball cap, or even a ring can take control of the motion of the mouse pointer on the screen.

7   GAME CONTROLLER

When the viewer's experience can take place inside a space where people have the sensation of walking, sitting, standing, and interacting as a group, the mouse and keyboard become less practical for a viewer navigating inside a virtual space that is projected on a room-size screen. The keyboard may require extra information to be pasted on the keys, which can be difficult for the viewer to see in the dark. In addition, the mouse needs a flat surface, which may be difficult if you are unable to anticipate when viewers are going to be standing, walking, or sitting. New challenges emerge for interactive designers in situations where viewers may develop hand, head, and feet gestures and head motions as a result of new relationships with the virtual environment and available input devices. The game controller pictured here can free the viewer by replacing the traditional mouse and keyboard. For example, the controller can be dangling from the ceiling with a long USB extension and can be used in a standing or seated position. The programmable keys of the controller are also very intuitive for the viewer.

8   SOUND AS AN INPUT DEVICE

A microphone is an input device that can be used in a very intuitive and natural way by the audience. The most inhibited viewers will feel comfortable clapping their hands, stamping their feet, or snapping their fingers. Shotgun microphones can be hidden behind the screen, or piezo microphones can be taped to the floor. Although the choice of the microphone is important depending on the size of the space and the distance from the viewer, only specific elements of the sound—for example, pitch and volume—will be processed to trigger actions in a virtual world.

Behaviors can process sound according to the value of a parameter. For example, the sound behavior can determine if the value of the volume reaches a certain threshold to trigger an action. In other cases, the sound behavior checks the variation of value of a sound parameter—for example, moving an object in a virtual space according to the volume of the sound input. In this case, the variation of value of the volume of the sound coming from a microphone can move an object.

8.1   Using a Threshold

IAGetVolBB is a building block created by Ronald Hof at www.sphaero.org/projects/ia and is available on the companion CD-ROM.

The game controller can be used in a standing or seated position. The programmable keys of the controller are very intuitive for the viewer.

IAGetVolBB can process the volume of the sound captured by a microphone. The interactive demo can be found on the companion CD-ROM in the Viewer's Experience folder under the Sound Analyzer subfolder.

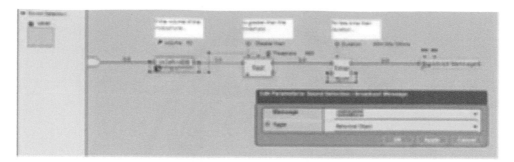

IAGetVolBB can process the volume of the sound captured by a microphone.

Values of the volume of the sound are compared to a threshold value—for example, 500. Values are tested against the threshold for a given time—for example, every 100 milliseconds. If the sound input is greater or equal to the threshold value during that time, a message, "Switch camera target," is broadcast inside the virtual world.

8.2   Sampling Sound

The asFFT building block, created by Antoine Schmitt, can process a variation of values of the volume of the sound captured by a microphone. The values are converted into variables that can be used in various ways. For example, a variable can control a translation movement or a color change. The following example shows how several particle emitters can move and change color depending on the volume of the sound input. The example shows how a particle emitter attached to 3D frames can change position and color according to variations of volume of sounds captured. The top view shows a virtual space changing motion and color according to variations of the sound input when someone is talking in front of a microphone. The interactive demo can be found on the companion CD-ROM in the Viewer's Experience folder under the AsSound 2.5 and 3.0 subfolder.

This sound example created by Antoine Schmitt uses a powerful Virtools feature called the VSL Editor. The Run VSL building block controls the position and motion of 3D frames. The code for the Run VSL building block, attached to the master 3D frame, is as follows:

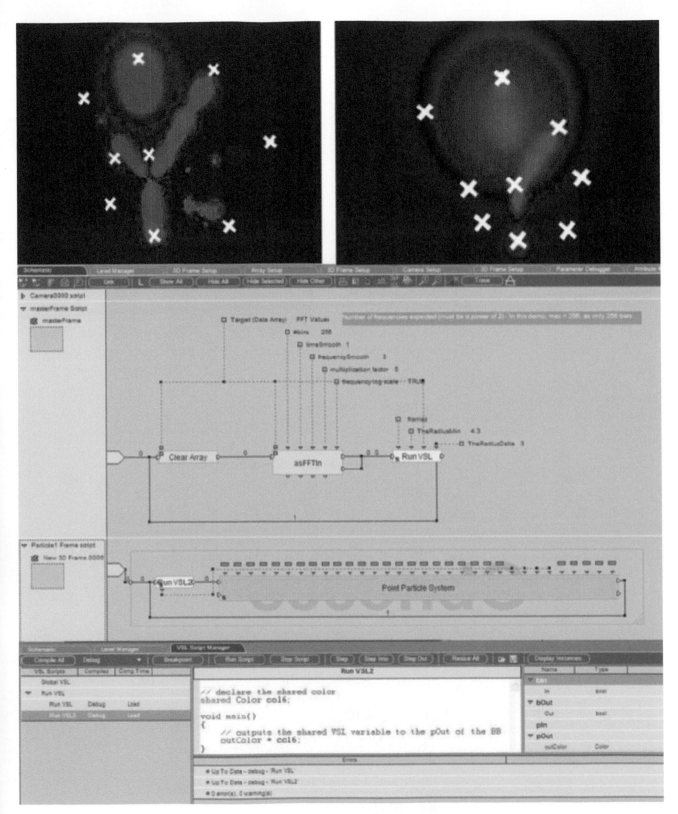

The top view shows a virtual space changing motion and color according to variations of the sound input when someone is talking in front of a microphone. The middle view shows behaviors for particle emitters attached to 3D frames. The bottom view shows coding in the VSL editor.

```
// Declare the shared color
shared Color col6;

void main( )
{
  // Get number of objects
  int nbObject = objGroup.GetObjectCount( );

  // Check fftValues array
  if (fftValues.GetColumnCount( ) < 1) {
    // not yet good
    bc.OutputToConsole("array not ready");
    return;
  }
  int nbValues = fftValues.GetRowCount( );
  if (nbValues < nbObject) {
    // drop it
    bc.OutputToConsole("not enough values");
    return;
  }

  // Calculate angle between each objets
  float angle = 2*pi/nbObject;

  Vector pos(0,0,0);
  // Place each object on the circle

  for (int i = 0; i < nbObject; ++i) {
    float fftValue;
```

```
// We get the nth object and cast it in Entity3D

Entity3D obj = Entity3D.Cast(objGroup.GetObject(i));

// If cast failed obj is null so we skip process...

if (!obj) continue;

// Get the values

fftValues.GetElementValue(i, 0, fftValue);

// Compute object position

float radius = radiusMin + 5.0*fftValue;

pos.x = radius*sin(i*angle);

pos.z = radius*cos(i*angle);

// Place object

obj.SetPosition(pos, null, true);

if (i == 6) {

    int r, g, b;

    // Greener if louder, redder if no sound

    r = 255.0*(1.0 - fftValue);

    g = 255.0*fftValue;

    b = 0;

    col6.Set(r, g, b, 0);

    }

  }

}
```

The Run VSL2 building block controls the color of the particle generator attached to the 3D frames located at the periphery of the wheel. The code for the Run VSL2 building block is as follows:

```
// Declare the shared color
shared Color col6;

void main( )
{
    // Outputs the shared VSL variable to the pOut of the BB
    outColor = col6;
}
```

9   NAVIGATION

Interface design adds a sense of context and helps viewers to understand how they can influence or control a virtual world. The interface design includes navigation systems to help viewers understand where they are and what they will do next in the virtual world.

From reading the previous sections of this chapter, you already have an idea of how to put together many different technologies and how they can exchange data with a virtual world. The technological path that connects the viewer's input device to a virtual world displayed on the screen is sometimes torturous. We expect every element of the path to play the right note even if they are very different. Although the elements of the technological path are diverse, there is an expectation for them to work seamlessly together. The technological path can be visualized as an ensemble of musicians playing so fast together, at least 30 frames per second, that the viewer hears only one note at a time. Navigation is a psychological element that takes place in the middle of hardware elements present on the path. Navigation occurs after the section on input devices and before the section on screen displays. In a way, navigating through a virtual world feels like following the advice "first do something and then deal with the consequences," which is known to be a source of problems in the real world.

The following three questions may be helpful to ask the viewer while designing a navigation system in a virtual world:

1 As a result of using the input device, can you tell if you are staying in one place watching the motion, transformation, evolution, and modification of the world around you or if are you moving through a virtual world?

Are you watching it from a static point of view or from a moving point of view? If you stay in one place, can you describe where you are and the changes that you witness? For example, textures may change color, objects may become transparent, or the lighting may change.

2 What gives you the feeling that you are moving? How does the input device influence the way you travel through space? For example, how do you perceive speed, acceleration, or deceleration? Describe what you see while traveling through space.

3 Why are you doing this? What is your goal? Did you receive instructions or guidance toward a goal? Is the experience bringing you something different than other media?

I suggest that you organize the questions while viewers experience the main scenes. Testing a three-scene prototype with viewers can give a good idea of the quality of the product being developed.

For each scene sequence ask viewers to describe what they see changing, what can be explored, the options or incentives given to them to explore more, how to achieve the goal, or how to go to another scene. Please keep in mind that giving too many options can confuse the viewer and generate unnecessary production work on your side.

9.1   Example of Process

The following questions can help to outline a discussion about navigation:

1 Describe the types of input devices to be used for interactive navigation. They may include, but not be exhausted by, the following:

- Mouse

- Keyboard

- Other controller such as a game pad

- Phone

- Sound

- P5 glove

- Others

2 Describe additional interactive parameters—for example, changing speed, changing color, translation, rotation, scale, and switching scripts. The answer can include the following:

The chopper flies on a path activated by a game controller.

Script one path-navigation with speed changes.

Script two free navigation with free translate, rotate finding the rescue team on a timer. When time is over, the mission is canceled and the chopper returns on path-script one.

3 Organize your questions about the viewer's experience in the main scenes. For example:

- The first scene is the opening of the world.

- Basic navigation takes place during the second scene.

- What can be discovered during the next scene?

The answer can be as follows:

- The viewer discovers the scene from the viewpoint of a pilot sitting in the chopper's cockpit. The opening scene camera targets the rescue team located on the mountain; the viewer also needs to click on a red button inside the cockpit to start descending toward the target. The chopper gradually loses altitude while going through a snow storm—emulated by particle animation—which reduces the visibility of the pilot. The chopper does not collide with the terrain of the mountain, but slides at a given distance from the terrain—where collision detection is activated with a radius of X.

9.2   Interface Design and Content Management

The viewer's interface can be static or dynamic. Now that we have looked at the role of interface for telling a story, we can look in detail at behaviors that will implement and complement the interface you create.

Static interfaces can be made of 2D images glued to the rendering window or to the frame of the camera. These images, called 2D sprites, are bitmaps that are snapped onto the frame of the camera following it wherever it goes. These 2D sprites can be scripted as interactive buttons. For example,

This scene from Nighthawks is using a static interface, a push-button 2D image in Step 1, and a dynamic interface, the 3D arrow following the character in Step 2.

buttons can react to the following mouse controls: Mouse Down, Mouse Click, Rollover, Double Click.

Dynamic interfaces can react or even anticipate the viewer's decision. For example, a dynamic 3D arrow can replace the mouse pointer, indicating changes of direction in three dimensions in addition to changes of speed. The 3D pointer helps to visualize the viewer's intentions. The length of the arrow indicates how fast and where the viewer is going. The same dynamic arrow can also be placed on top of a virtual character. The presence of a 3D arrow on top of a virtual character inside a crowd indicates that a viewer picked up the character. The direction of the 3D arrow indicates where the viewer chooses to move the virtual character. The 3D arrow turns into a yellow sphere when the character is idle. The yellow sphere fades out when the viewer no longer controls the character.

Let's look at the scripts and behaviors that can activate static and dynamic interfaces.

9.2.1 Static Interfaces—Creating a Push Button and a Mouse Rollover

A static interface is made of 2D images that are always displayed in the same place on the screen. The images on the screen are scripted with a PushButton building block. The 2D sprite waits for a mouse click to trigger a behavior.

An example showing several ways to trigger events with the PushButton building block is the Sequencer building block that can switch between two or more behaviors for every mouse click. In Steps 4 and 5, the Sequencer is replaced by a message system. In this case, the Send Message behavior activates another script, Steps 6 and 7. A step-by-step tutorial on setting up buttons is available on the companion CD-ROM in the folder for Chapter 8.

Let's look at how a mouse rollover works. When the mouse pointer covers an image on the screen, the image is detected by the Mouse Waiter building block. The Switch on Parameter building block activates the corresponding pin number for the image. The images with matching numbers

In this example of a dynamic interface, the presence of a 3D arrow on top of a virtual character inside a crowd indicates that a viewer picked up the character. The direction of the 3D arrow indicates where the viewer wants to move the virtual character.

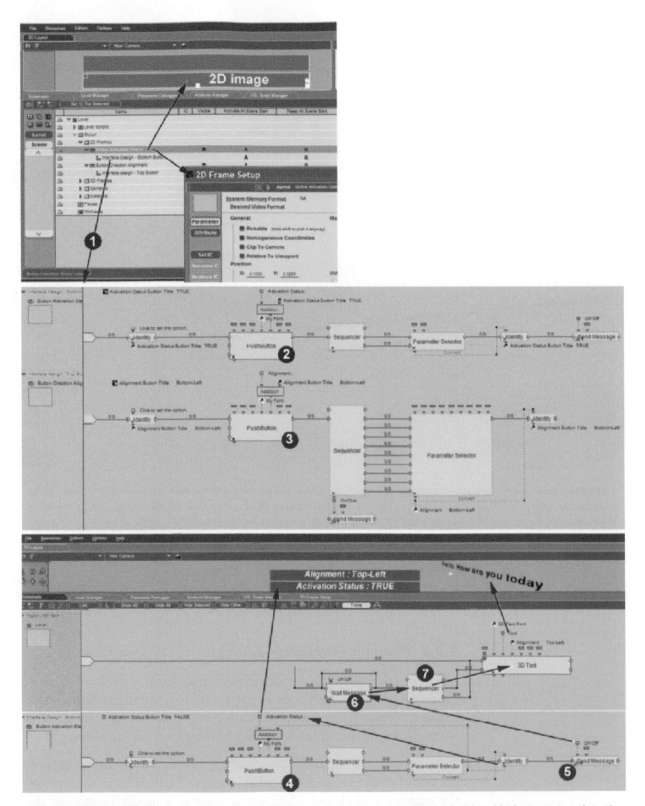

Step 1: The images on the 3D Layout screen are 2D frames scripted with a PushButton building block. Steps 2 and 3: The Sequencer building block can switch between two behaviors for every mouse click. A sequencer with multiple outputs allows a viewer to control several behaviors. Step 4: The Send Message behavior, which is part of a message system, replaces the sequencer and activates another script, Step 5 and 6. This string of behaviors can also include a Sequencer building block, Step 7.

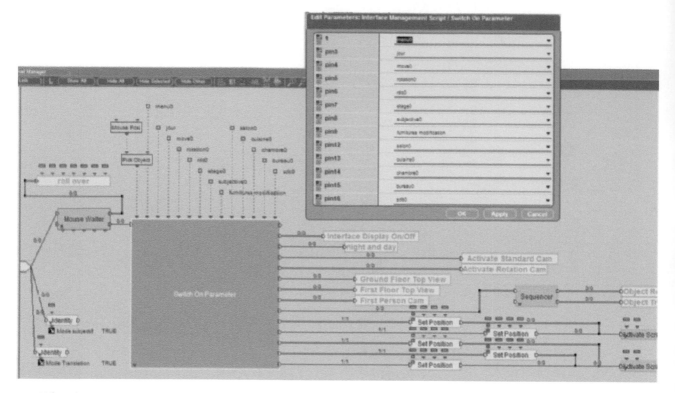

When the mouse pointer covers an image on the screen, the image is detected by the Mouse Waiter building block. In this example, the Switch on Parameter building block activates the corresponding pin number for the image and for the script associated with the images. Images with matching numbers are on the top of the building block and the scripts are on the right of the building block.

are on the top of the building block, and the scripts are on the right of the building block.

9.2.2 Dynamic Interfaces—Examples of 3D Interfaces

Following is an illustration of dynamic interfaces that show how to use a dynamic 3D arrow as a 3D mouse pointer moving through a virtual space and as a 3D interface for a virtual character. In both cases, the 3D arrow is controlled by mouse inputs or by keyboard inputs from the viewer.

The 3D arrow is a navigation tool that informs the viewer about the ways she or he is moving through a virtual space. The yellow 3D arrow replaces the traditional black and white arrow, used as a mouse pointer on the

Examples of dynamic interfaces using a dynamic 3D arrow. In the left view, a 3D arrow displayed on top of a character indicates the walking direction. In the right view, a 3D mouse pointer is used to look at architectural details. In both cases, the 3D arrow is controlled by a mouse input or by a keyboard input.

screen, and indicates which direction the viewer is moving inside the virtual space. The 3D arrow indicates motion in four directions in addition to the speed of motion. The mouse pointer turns into a sphere when the mouse is idle.

9.2.3 How to Create a 3D Mouse Pointer

The Get Mouse Scale, Set Angle, and Set Position behaviors control a vector that transforms the position and appearance of a 3D mouse cursor. The interactive demo can be found on the companion CD-ROM in the Viewer's Experience folder under the 3dArrow subfolder.

The second script takes care of the 3D mouse movements through the virtual space. Forward-Backward, Up and Down, and Left and Right are behaviors that are used for both the keyboard input and the mouse input.

* Step 1 shows the setup for the keyboard with keys controlling Forward-Backward, "Up and Down, and Left and Right".

* Step 2 shows the Mouse Controls behavior can be broken down into several subroutines.

* Steps 3, 4, and 5 show the details of each of the subroutines.

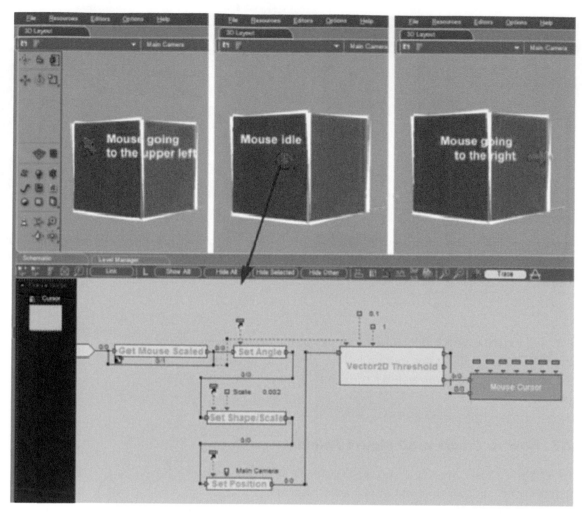

This illustration shows how to create a 3D mouse pointer. In the top view, the 3D arrow shows how the viewer is moving in the virtual space. The mouse pointer turns into a sphere when the mouse is idle. In the bottom view, the Get Mouse Scale, Set Angle, and Set Position behaviors control a vector that transforms the position and appearance of a 3D mouse cursor.

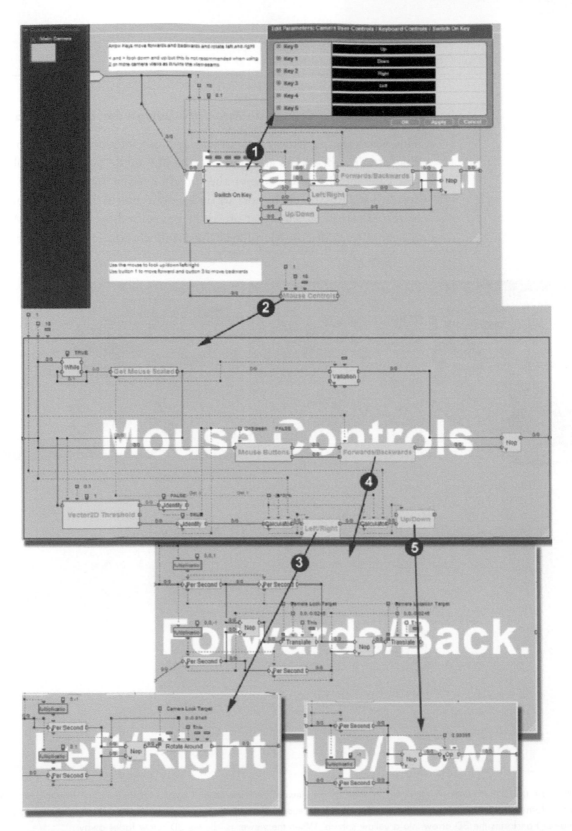

Forward-Backward, Up and Down, and Left and Right are behaviors that are used for both the keyboard input and the mouse input. Step 1 shows the setup for the keyboard with keys controlling Forward-Backward, Up and Down, and Left and Right. Step 2 shows the Mouse Controls behavior exploded into several subroutines. Steps 3, 4, and 5 show the details of each of the subroutines.

9.2.4 *How to Create an Interface Design for a 3D Character*

This next example shows how to display a 3D interface on the top of a character. The behaviors are similar to the 3D mouse pointer example. In this case the 3D arrow shows the direction given by the viewer. The Forward and Rotation behaviors are designed to control the transformations of the 3D arrow in ways that are easily recognized by the viewer. When the character is idle, the Morphing behavior slowly transforms the 3D arrow

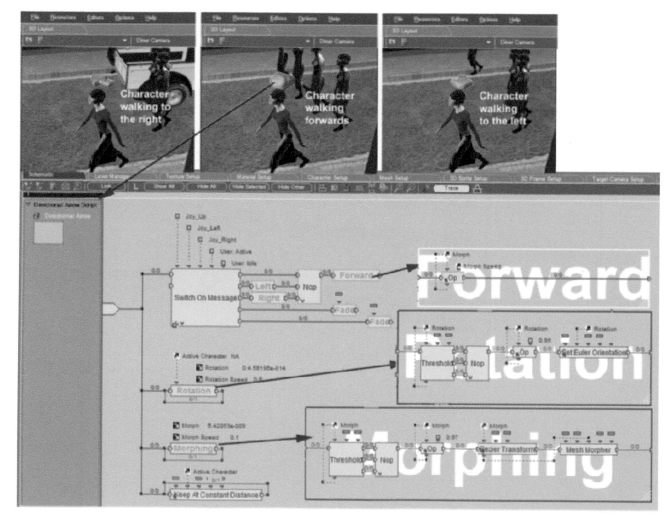

In this case the 3D arrow shows the direction given by the viewer. The Forward and Rotation behaviors are designed to control the transformations of the 3D arrow in ways that are easily recognized by the viewer. When the character is idle, the Morphing behavior transforms the 3D arrow into a yellow sphere. When the viewer is idle the 3D arrow fades away.

into a yellow sphere following the smooth curve of the Bezier Transform building block. When the viewer is idle the 3D arrow fades away.

10   DISPLAYS

10.1   Multiscreen Displays

Displays are the last elements on the technological path covered in this chapter. From flat-screen displays, to video projections inside a "cave" environment, to Times Square's giant urban screens, virtual spaces can be displayed in many different ways in space. Brightness is always an issue for screens and projectors because it can limit the use of daylight screens.

The quality of the interaction with a virtual space depends more on the position of the screen in space than on the screen size. The relationship with the viewer depends on the way a virtual space is projected onto a screen hanging on a wall or on a flat surface such as a table. In the case of a virtual surgery project, the surgeon can touch the image of a virtual patient projected on a flat table. A virtual hand appears on top of the image of the virtual body following the surgeon's hand. A motion-tracking device detects the motion of the surgeon's hand and sends data to the virtual hand.

The rendering window can display one or several frames coming from virtual cameras inside the virtual world. Current video card technology allows one 3D acceleration card to be connected to three displays. Please keep in mind that the rendering process taking place in the 3D acceleration card is divided among the number of additional views. The overall speed of a setup with multiple screens is roughly divided by the number of views included in the display. The overall performance is also dependent on the number of displays. In some cases, the full-screen mode may not be available across three monitors. The Matrox Parahelia card, illustrated here, can display a virtual world on three contiguous screens.

Let's look at ways to program behaviors for the rendering window. The behaviors described in this section include the following:

- Split views that control the layout of windows receiving images from two or three virtual cameras. This behavior can automatically update the choice of the active virtual camera.

- Multiple screens display seamless images coming from two or three cameras.

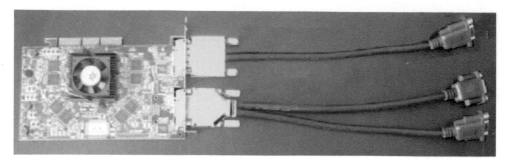

The Matrox Parahelia card can display a virtual world on three contiguous screens.

10.2   Split View with Two Rendering Windows

The following example uses a two-monitor setup with images coming from two virtual cameras. The Resize View building block, Step 1, defines the dimensions of the rendering frame for the main camera. The Additional View building block, Step 2, creates an additional frame for the second camera. The dividing line between View 1 and View 2 matches with the edge of the two flat-screen monitors. The 3D acceleration card, Step 3, has two VGA outputs. Steps 4 and 5 have one output for each monitor or video projector.

10.3   Comics Split Views

Creative ways of dividing the real estate of a screen can be inspired by comics. The screen can be split into several views, similar to the comic page layout in a newspaper, where each frame is a different viewpoint of the story. Although low-cost display technology may not be advanced enough to fully take advantage of these behaviors, expensive virtual reality packs with rendering distributed to multiple computers are already available. (They are generally referred to as clusters.) We can imagine that a 3D acceleration card with six or more VGA outputs will be available in a near future.

The example on page 402 shows how to set up a flexible layout screen that can be split into three views. To display three screens, we will reuse the Split View with Two Screens behavior with one more Additional View building block. This setup has only been tested on a single monitor.

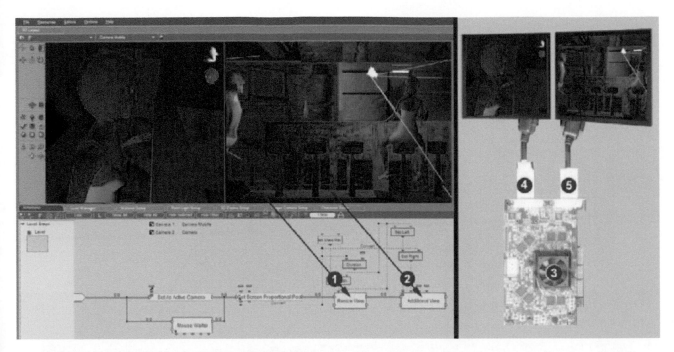

The Resize View building block, Step 1, defines the dimensions of the rendering frame for the main camera. The Additional View building block, Step 2, creates an additional frame for the second camera. The dividing line between View 1 and View 2 matches with the edge of the two flat-screen monitors. The 3D acceleration card, Step 3, has two VGA outputs. Steps 4 and 5 have one output for each monitor or video projector.

10.4   Multiple Screens Display

A three-screen display can be created from a single camera with the proper depth of field or with a node of three cameras. In the case of using multiple cameras, the views are contiguous because they are filmed by a node of three cameras attached together inside the virtual world. The setup is similar to holding a node of three camcorders rigged together while walking in a space.

The 1-2-3 Screens behavior allows the user to toggle between one-screen, two-screen, or three-screen displays by using keys 1, 2, and 3 on the keyboard. The 1-2-3 Screens behavior allows three contiguous views from a node of three contiguous virtual cameras inside the virtual world. The "cave" effect provided by the group of cameras is the equivalent of three views projected on the right, front, and left walls of a room.

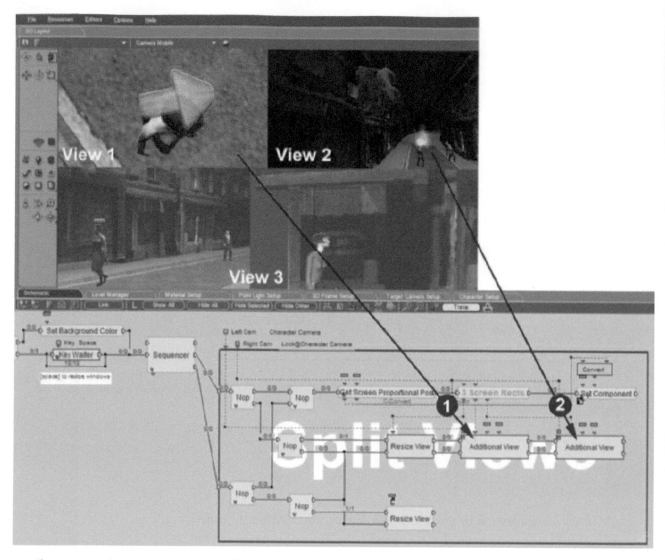

The screen can be split into several views like a comic page layout where each frame is a different viewpoint of the story. View 3 is associated with the main camera. Steps 1 and 2 show the additional views.

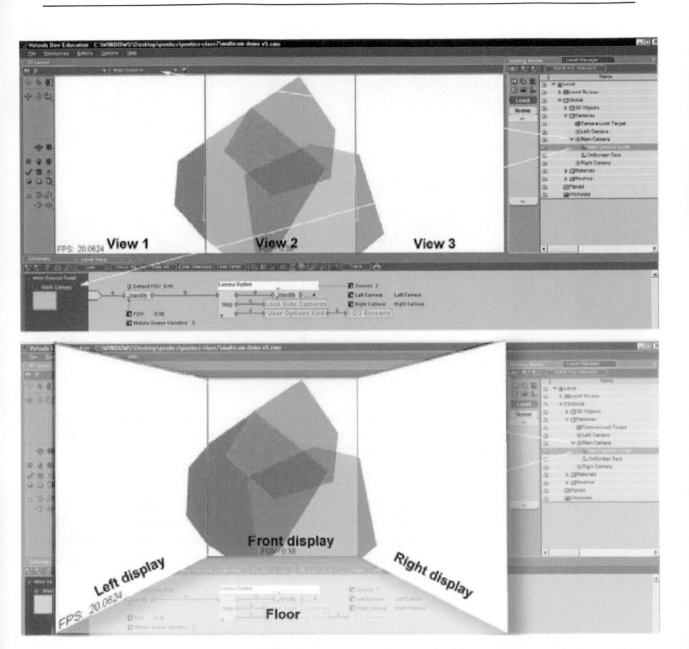

In the top view, the 1-2-3 Screens behavior allows three contiguous views from a node of three contiguous virtual cameras inside the virtual world. In the bottom view, the "cave" effect provided by the group of cameras is the equivalent of three views projected on the right, front, and left walls of a room.

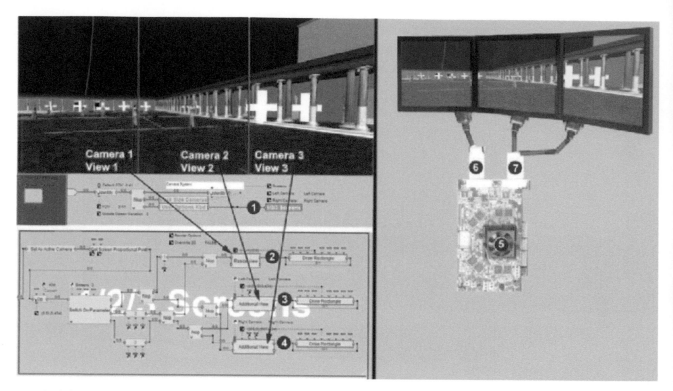

The behavior added to a main camera script will automatically divide the screen on your desktop into three views (Step 1). The behavior can resize the screen to the width and height of the desktop (Steps 2 through 4). The video card ultimately broadcasts three VGA signals, one for each monitor or video projector (Steps 5 through 7).

The 1-2-3 Screens behavior added to a main camera script will automatically divide the screen on your desktop into three views (Step 1). The behavior can resize the screen to the width and height of the desktop (Steps 2 through 4). The video card ultimately broadcasts three VGA signals, one for each monitor or video projector, Steps 5 through 7.

11   CONCLUSION: THE SHAPE OF THINGS TO COME

The New York Cube is a total immersion experience that explores new directions in the use of input devices and multiscreen displays. The Cube is a light metallic structure that can be suspended from a larger structure, for example the roof of a hangar. Visitors are attracted by the object floating above the ground. The cube opens its motorized panels when someone walks close to its faces. Vector images start to scroll on the faces of the cube,

Illustration for the New York Cube, a total immersion experience created by Jean-Marc Gauthier, Miro Kirov, and James Tunick. Visitors are now standing in the middle of a virtual museum, high up in the sky above the streets of Manhattan. A gesture of the hand and the space starts to move slowly following the direction of the hand.

textures appear, the sound gets clearer, and a virtual space can be seen moving inside the cube. Visitors are now standing in the middle of a virtual museum, high up in the sky above the streets of Manhattan. A gesture of the hand and the space starts to move slowly following the direction of the hand. Snapping fingers move the space faster. Virtual people from New York appear on the screens and acknowledge the visitors. Navigation messages scroll on the floor.

The cube allows people to experience new kinds of relationships with the rest of the world. The virtual spaces presented inside the cube allow visitors

- to be immersed inside the four dimensions of a virtual space and to play with behavioral engines controlling several media: visuals, music, voice, and animation.

- to be part of live shows taking place on the "reactive" stage inside the cube.

- to communicate with other groups of people in the world through telepresence. The setup allows artists to create artwork remotely, with

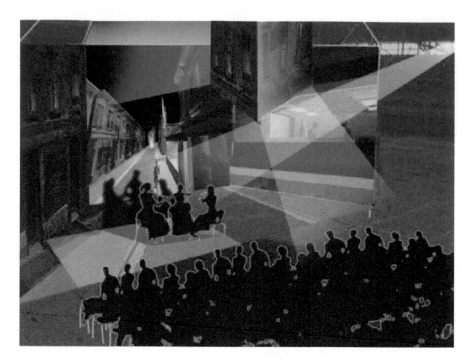

This illustration shows a jazz performance taking place inside the cube. A digital network made of motion sensors, behavioral engines, spatial sound, and three-dimensional graphics can interact in real time with the musicians and the public. The New York Cube is a total immersion experience created by Jean-Marc Gauthier, Miro Kirov, and James Tunick.

View of the outside of the New York Cube showing the screens, the reflective mirrors, and the video projectors. The New York Cube is a total immersion experience created by Jean-Marc Gauthier, Miro Kirov, and James Tunick.

meetings and collaborations between visitors and artists taking place inside the cube.

The design of this project shows how a digital world can become part of the architectural space where we live, similar to the concepts played with in *Star Trek*'s "Holodek". This environment is created on top of a digital network made of motion sensors, behavioral engines, spatial sound, and three-dimensional graphics.

The virtual actors and other objects inside the virtual space have received their scripts, but the technological path brings interactivity, inflections, tempo, and motion to all of the elements of the virtual world. The interface is designed like a map of behaviors where, in real-time, the viewer decides what is possible and where to go based on their instinct and without any predefined map.

Bibliography

The following list includes books, games, and movies that were helpful in the writing of this book.

Alekan, H. (2001). *Des Lumieres et des Ombres*. Paris: Editions du Collectionneur.

Alton, J. (1955). *Painting with Light*. Berkeley: University of California Press.

Antonioni, M. (1960). *L'Avventura*. Movie. The camera exploring the landscape of the island has a physical presence similar to one of the characters on the island.

Cavell, S. (1981). *Pursuit of Happiness—The Hollywood Comedy of Remarriage*. Harvard: Harvard Film Studies.

Champandard, A. (2003). *AI Game Development—Synthetic Creatures with Learning and Reactive Behaviors*. Indianapolis: New Riders.

Christy, D. (2000). *Luminosity: The Paintings of Stephen Hannock*. San Francisco: Chronicle Books.

Eberly, D. (2004). *Game Physics*. San Francisco: Morgan Kaufmann Publishers.

Ettedgui, P. (1999). *Production Design and Art Direction Screencraft*. Woburn: Focal Press.

Fleischer, R. (1966). *Fantastic Voyage*. This movie recreates organs, flux, physics, and textures inside a human body. A group of scientists surrounding the patient in a lab communicates with a miniaturized medical team navigating inside the human body. This virtual visit of a human body is a precursor of many elements of the language of virtual spaces.

Franklin, S. (2001). *Artificial Minds*. Cambridge: MIT Press.

Gauthier, JM. (2002). *Creating Interactive 3-D Actors and their Worlds*. San Francisco: Morgan Kaufmann Publishers.

Gibilisco, S. (2002). *Concise Encyclopedia of Robotics*. New York: McGraw-Hill.

Harryhausen, R., & Dalton, T. (2004). *Ray Harryhausen.* New York: Billboard Books.

He L. (1996). *The Virtual Cinematographer: A Paradigm for Automatic Real-Time Camera Control and Directing.* AMC papers.

Hitchcock, A. (1954). *Rear Window.* Universal Pictures, producer. Movie.

Katz, S. (1991). *Film Directing Shot by Shot: Visualizing from Concept to Screen.* Studio City: Michael Wiese Productions.

Keller, M. (1999). *Light Fantastic: The Art Design of Stage Lighting.* Munich: Prestel.

Kilgard, F. (2003). *The CG Tutorial.* Boston: Pearson Education, Inc.

Kitchen, S. (1997). *Real World Bryce 2.* Berkeley: Peachpit Press

Konami. (2002). *The Making of Metal Gear Solid 2.* Tokyo: Konami Computers Entertainment.

Kubrick, S. (1957). *Paths of Glory.* James Harris, producer. Movie. Several examples of path cameras reveal physical relationships between the soldiers and the topography of the battlefield.

Kubrick, S. (1968). *2001: A Space Odyssey.* Movie. Cameras acting as an autonomous character can film beyond the life of one of the astronauts. This film is a constant source of inspiration for designing new virtual spaces and virtual cameras.

Kubrick, S. (1980). *The Shining.* Stanley Kubrick, producer. Movie. Steady cameras filming the chase inside the maze are in the center of a spatial vortex that can swallow the space surrounding the action.

Lecat, J.G., & Todd, A. (2003). *The Open Circle.* London: Faber and Faber Ltd.

Mason, M. (2001). *Mechanics of Robotic Manipulation.* Cambridge: MIT Press.

Melville, J.P. (1972). *Le Cercle Rouge.* Robert Dorfman, producer. Movie. Melville's film is a great source of inspiration for designing virtual spaces. The planning of the action inside the space and the visual exploration of the space of the action are great sources of inspiration.

Melville, J.P. (1972). *Un Flic.* Robert Dorfman, producer. Movie.

Molyneux, P. *Black and White*. Lionhead Studios. Game.

Nehaniv, C. (2002). *Imitation in Animals and Artifacts*. Cambridge: MIT Press.

Neumann, D. (1999). *Film Architecture: Set Designs from Metropolis to Blade Runner*. Munich: Prestel.

Rabin, S. (2004). *AI Game Programming Wisdom 2*. Hingham: Charles River Media.

Ray, N. (1980). *We Can Go Home Again*. Movie. Ray's clip inserted inside Wim Wenders movie *"Lightning Over Water"* is an example of a multiscreen movie where the viewer cannot see it all.

Rieser, M., & Zapp, A. (2002). *New Screen Media Cinema/Art/Narrative*. London: British Film Institute.

Rybczynski Z. (1990). *The Orchestra*. This video is exploring relationships between music, time, and virtual space.

Saunders, P. (2001). *Myst 3 Exile*. Presto Studios. Game.

Schafer, T. (1998). *Grim Fandango*. Lucas Art. Game.

Schuiten and Peters (2002). *Le Guide des Cites*. Paris: Casterman. This book presents examples of projects combining commix, installations, and virtual cities.

Scott, R. (1982). *Blade Runner*. Sir Run Run Shaw Thru, producer. Movie. The reference for the futuristic city created in the chapter on Textures.

Shaw, J. (2003). *Future Cinema—The Cinematic Imaginary after Film*. Cambridge: MIT Press. This book covers most of the research and experiments on the future of cinema.

Sokal, B. (2002) *Syberia*. The Adventure Company. Game created by a comics designer with Virtools.

Spielberg, S. (1984). *Indiana Jones and the Temple of Doom*. Lucas Films Ltd, producer. Movie. The race inside the mine is a brilliant illustration of associations between the narration and runaway mining carts.

Spielberg, S. (2002). *AI Artificial Intelligence*. Steven Spielberg and Stanley Kubrick, producers. Movie. The film, co-written with S. Kubrick, blends studio scenes and virtual sets.

Strafford, B. (2001). *Devices of Wonder*. Los Angeles: Getty Publications. This book shows early experiments of screen projections and viewer's interactions.

Thomsen, W. (1994). *Visionary Architecture from Babylon to Virtual Reality*. Munich: Prestel.

Thor, A. (2003). *Massively Multiplayer Game Development*. Hingham: Charles River Media.

Trappl, R., & Paolo Petta, S. P. (2002). *Emotions in Human and Artifacts*. Cambridge: MIT Press. Robotics research showing relationships between kinematics and expressions can be used for virtual characters.

Trauner A. (1988). *Decors de Cinema*. Paris: Flammarion.

Walker, A. (1972). *Stanley Kubrick Director*. New York: Norton and Company.

Webb, B. (2001). *Biorobotics: Methods and Applications*. Cambridge, MA: MIT Press.

Welles, O. (1958). *Touch of Evil*. Albert Zugsmith, producer. Movie. The opening sequence of the movie is a reference for designing autonomous virtual cameras inside virtual spaces.

Wenders, W. (1980). *Lightning Over Water*. Road Movies Filmproduktion GMBH. Movie.

Index

Printed and bound by CPI Group (UK) Ltd, Croydon, CR0 4YY

21/10/2024

01777094-0014